PROGRAMMING METHODS IN STRUCTURAL DESIGN

N. G. R. Iyengar
Professor of Aeronautical Engineering
Indian Institute of Technology
Kanpur

and

S. K. Gupta
Assistant Professor of Mathematics
Indian Institute of Technology
Kanpur

In collaboration with N. C. Nigam (Chapters 6 and 17),
Professor of Aeronautical Engineering,
Indian Institute of Technology, Kanpur

A HALSTED PRESS BOOK

JOHN WILEY & SONS
New York

First published in the United Kingdom in 1981 by
Edward Arnold (Publishers) Limited
Published in the U.S.A.
by Halsted Press, a division of
John Wiley & Sons, Inc., New York.

British Library Cataloging in Publication Data

Iyengar, N. G. R.
  Programming methods in structural design.
  1. Structural design—Data processing
  2. Programming (Electronic computers)
  I. Title   II. Gupta, S. K.   III. Nigam, N. C.
  624.1′771′028542   TA658.2

ISBN 0–470–27298–8

Printed and bound in Great Britain
at The Pitman Press, Bath

Preface

Of all feasible designs for a structure, those that can resist the loads imposed on it and are aesthetically appealing are considered the most suitable. Since several such designs exist for most structures, it is desirable to choose the most economical one. For example, economy can be effected in aeronautical structural design by keeping at the minimum the weight, which is directly proportional to the cost. If cost economy of even a small order can be brought about, the total saving can be sizable. The incorporation of optimization concepts when designing a structure invariably results in improved techniques. In this text, we discuss such design techniques and apply them to obtain an optimum structural design.

The volume is divided into four parts—Part I (Chapter 1), Part II (Chapters 2–6), Part III (Chapters 7–10), and Part IV (Chapters 11–17). Chapter 1 explains the need for optimization techniques in deterministic and probabilistic design. Chapters 2–5 deal with the static and dynamic analysis of beams and plates and the numerical techniques that can be effectively used on a computer. Chapter 6 is devoted to random vibration principles. Chapters 7–10 cover various optimization techniques, the stress being on the techniques rather than on developing them mathematically. The discussion here is intended primarily as background material; as such, it includes only those aspects of optimization techniques that are necessary for a proper understanding of their different applications to structural designs, which are exhaustively treated in Part IV. In this final section (Chapters 11–17), the programming techniques are applied to, among others, the design of aircraft wing structures, box beams subjected to various types of loading, and the design of roofs. A number of optimization problems considered in random vibration environment are also included. In addition, programming techniques are applied to obtain eigenvalues and eigenvectors for nonlinear vibration problems. The numerical calculations have been worked out on the IBM 7044 computer and the flowcharts used in developing the programs, here and elsewhere in the text, are also given.

The book has resulted from lectures delivered at the Indian Institute of Technology, Kanpur, and from a QIP course on programming methods in engineering design conducted at the Institute. It is intended for undergraduate and postgraduate engineering students and also for practising engineers.

We are extremely thankful to Professors P. N. Murthy and S. R. Adidam for permission to use some of the results they have obtained. The financial assistance from the Educational Development Centre at the Indian Institute of Technology, Kanpur, towards the preparation of the manuscript is also gratefully acknowledged.

Kanpur,
June 1980

N. G. R. Iyengar
S. K. Gupta

Contents

Notation

Structural analysis and applications

A_b	area of cross-section of beam
A_c	area of cross-section of column
D	flexural rigidity of plate
D_x, D_y	flexural rigidity of plate along X-, Y-direction
E	elastic modulus
e	strain vector
f	natural frequency (cycles/sec)
G	shear modulus
$[G]$	total geometric stiffness matrix
$[g]$	element geometric stiffness matrix
I	second moment of area
I_f	second moment of area of flange
I_0	polar moment of inertia about shear centre
J	St. Venant torsional constant
K, k	buckling coefficient
$[K]$	total elastic stiffness matrix
$[k]$	element elastic stiffness matrix
L	length of column
l	length of element
M	internal resistive bending moment
$[M]$	total mass matrix
M_x, M_y, M_{xy}	bending and twisting moments per unit length of plate
$[m]$	element mass matrix
m_b	mass of beam per unit length
m_p	mass of plate per unit area
N	number of elements

$[N]$	matrix of shape functions
N_x, N_y, N_{xy}	inplane normal and shear force per unit length
$(N_x)_{cr}$	critical load per unit length
P_{cr}	critical load for column
Q_x, Q_y	transverse shear force along X-, Y-direction
q	nodal displacement vector
S_x	shear load along X-direction
T	kinetic energy
T_J	St. Venant torque
T_Γ	torsion-bending torque
U	strain energy
U_b	strain energy due to bending
u	displacement along X-direction
V	potential energy
v	displacement along Y-direction
W_d	work done by applied load
w	displacement along Z-direction
X	cartesian coordinate for beam and plate
X_x, Y_y, Z_z	prescribed body force per unit volume
x_0	distance along X-direction of shear centre from centre of gravity
Y	cartesian coordinate for beam and plate
y_0	distance along Y-direction of shear centre from centre of gravity
Z	cartesian coordinate for beam and plate
β	twist of column
$\epsilon_x, \epsilon_y, \epsilon_z$	normal strain along X-, Y-, Z-direction
λ	eigenvalue
ν	Poisson's ratio
ξ, η	nondimensional parameters
π_p	total potential energy
ρ	radius of gyration
ρ_b	density of beam material

ρ_p	density of plate material
σ	stress vector
Φ_i	prescribed surface forces per unit area
$[\chi]$	material matrix
ψ	bending slope
ω	natural frequency (rad/sec)

Optimization techniques

$0_{p,q}$	null matrix of order $(p \times q)$
$A(\alpha)$	set depending on α
a_{ij}	(i, j)-th element of matrix A
b_i	i-th element of vector b
c_j	j-th element of vector C
$d(i)$	decision variable at stage i
F_n	n-th Fibonacci number
$f(X)$	value of function f at X
$f_N[X]$	optimum value of n stages with initial state X
G_j	state transformation function for j-th state variable
$g_i(X)$	value of constraint function g_i at X
H	Hessian matrix
H_k	approximation at k-th iteration of H^{-1}
$|H_k|$	determinant of H_k
I	system return function
I_p	identity matrix of order p
$\hat{M}$	index set of binding constraints
$\hat{N}$	index set of binding non-negativity restrictions
$P(X, r)$	function to be minimized in penalty function approach
R	set of real numbers
R^n	n-dimensional Euclidean space
$R(i)$	return function at stage i

X	element of R^n
X_B	basic components of vector X corresponding to basic matrix B
$(\hat{X}, \hat{U})$	saddle point
X^i	i-th point generated by algorithm
$X(r_k)$	point at which $P(X, r_k)$ is minimum
$x_j(i)$	j-th state variable at stage i
Y_{ij}	(i, j)-th element of matrix Y
Δa_{ij}	change in a_{ij}
Δb_{ij}	change in b_{lj}
Δc_{ij}	change in c_{ij}
ϵ	tolerance limit
λ	scalar
$\Phi(X, U)$	Lagrange function
$\nabla f(x)$	gradient vector of function f at x
Roman capital letters	sets or matrices

PART I   GENERAL

1

Introduction

1.1 OPTIMUM DESIGN

Decision-making features predominantly in all fields of human activity, including scientific and technological, and affects every sphere of life. Engineering design, which entails sizing, dimensioning, and detailed elemental planning, is not exempt from its influence. For example, an aircraft wing can be made from aluminium or steel and, once the material and shape are chosen, there are many ways of devising the required internal structure. Again, in civil engineering, designing a roof to cover a large area devoid of intermediate columns requires either T-beams or grid structures. Within such a design framework, there are variables that specify the dimensions, proportions, and other details of the structural design.

The ultimate aim of the design process is to improve an existing design or replace an unacceptable design by one that meets the given requirements, that is, to choose the best possible design which satisfies all the restrictions placed on it. This is called the *optimum design*. It is arrived at by optimization techniques which play an important role in engineering design. In fact, optimization is inherent in the very process of designing.

The decisions leading to the selection of the optimum design may be intuitive (indirect) or logical (direct).

1.1.1 INDIRECT DESIGN

Indirect (or *intuitive*) *design* is based on past experience or random selection. Once a decision has been taken, an attempt is made to evaluate the consequences. This is the crux of indirect design.

The various stages of indirect design can be best understood by considering a specific example. In designing a bridge, the number of spans, the materials, shape, and cross-sectional dimensions can be intuitively selected. An analysis is then carried out to determine whether the *design constraints* are satisfied. These constraints usually limit such behavioral quantities as stresses, deflections, natural frequencies, and buckling loads on various components, and some of them might be violated because of the intuitive approach to the design. Often, the design analysis has to be repeated several times before the final design is obtained. This is a major drawback of indirect design. Another disadvantage is that even if a feasible solution is found, it cannot be claimed to be the best.

1.1.2 DIRECT DESIGN

Direct (or *logical*) *design* overcomes some of the drawbacks of indirect design. The basic feature of this design process is that it entails only logical decisions. In this method, we first establish all the behaviour constraints (on, for example, stresses and deflections) and the side constraints (for instance, on physical dimensions) and then maximize or minimize a function called the *objective function* (i.e., cost or merit function). For instance, when designing a bridge, this function is the total cost of the bridge which includes the direct cost (e.g., of materials) as well as the indirect cost (e.g., of shuttering). However, in designing structures such as aircraft and rockets, the objective function is the total weight of the structure since, in flying a heavy aircraft and launching a heavy rocket, weight is of prime importance. The objective function therefore covers a wide range of considerations, some of which can be expressed in a well-defined functional form.

1.2 AUTOMATED DESIGN

In the last two decades, remarkable progress has been made in the theory of structural design. Engineers and scientists have been under great pressure to design the best possible structure in the shortest time. The increasing use of computers in structural analysis has led to the concept of *automated design* which considerably reduces the time for each design step. In the initial stage, it is customary to base a structural design primarily on the requirements of strength and stiffness. Then the dynamic requirements are satisfied by a process of trial and error. With the availability of algorithms and sophisticated computers, attempts are being made to include all possible constraints and to achieve an optimum design. In such an optimum design, automated analysis is a component part of the overall design process (see Fig. 1.1).

The shortcoming of the automated optimum design is that, as with indirect design, considerable time is expended on the repeat analysis. Efforts are therefore being made to develop time-saving, yet accurate, efficient computational procedures.

1.3 RELIABILITY-BASED DESIGN

Structural optimum design methods can be classified according to the design concept used. Most civil engineering structures are still evolved on the basis of design constraints representing a permissible stress criterion. Some of the recent methods use a specified factor of safety against failure of the structure. This approach, referred to as the deterministic approach, has a number of limitations. For example, while designing a tall building or an antenna tower, wind loads are not deterministic. Here, as in most engineering design problems, the reliability-based design provides a more realistic treatment of the inherent uncertainty. In such a treatment, the objective function is expressed on the basis of probability theory.

1.4 PROGRAMMING METHODS

Programming methods deal with solution procedures for a certain class of optimization problems, and especially with the techniques of *linear programming*, *nonlinear programming*, *geometric programming*, and *dynamic programming*. Depending on the type of problem, we can directly or indirectly apply one or more of these techniques.

Very few structural engineering problems can be grouped under linear programming problems since the objective function and/or the constraints are nonlinear functions of design variables. However, it is sometimes possible to cast a nonlinear programming (NLP) problem into an approximately equivalent linear programming problem (LPP).

Another technique that can be used for structural optimization is geometric programming (GP). This technique does not attempt a direct solution of a problem, but, by a series of mathematical transformations, sets up a dual problem in which the constraints are linear. The dual problem is then solved by using the linear programming technique. Geometric programming is best suited for component optimum design where the number of parameters is

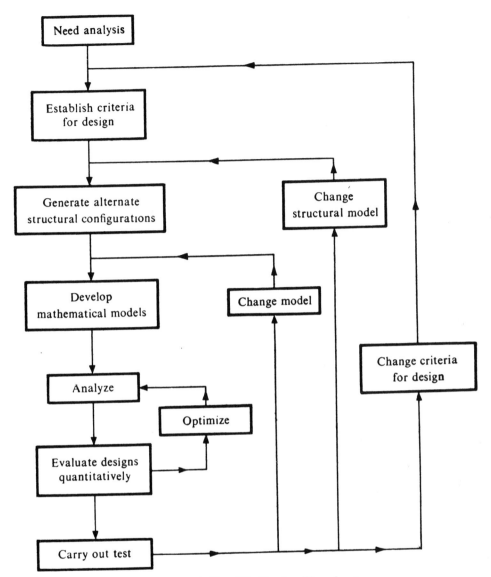

Fig. 1.1 A Possible Design Flowchart.

small in comparison with that of the total structure. However, all structural engineering problems cannot be solved by this technique.

The technique of dynamic programming is apt for solving problems in which the objective function, as also the constraint functions, are separable. In addition, if the number of constraints is small, substantial computational advantages accrue.

As already stated, optimization is inherent in the design procedure. It would, however, be futile to strive for optimization by using sophisticated mathematical tools when the analysis of the design—the vital factor—is only approximate. In the subsequent chapters of this text, therefore, the prerequisites to such analysis are covered in some length (Part II); relevant optimization techniques are then explained (Part III) and, finally, the analysis is carried out and optimization techniques applied to certain structural design problems (Part IV).

BIBLIOGRAPHY

Bellman, R. and Dreyfus, S. E., Applied Dynamic Programming, Princeton University Press, Princeton, 1962.

Duffin, R. J., Peterson, E. L., and Zener, C., Geometric Programming: Theory and applications, Wiley, New York, 1967.

Fiacco, A. V. and McCormick, G. P., Nonlinear Programming: Sequential unconstrained minimization techniques, Wiley, New York, 1968.

Fox, R. L., Optimization Methods for Engineering Design, Addison-Wesley, Reading, Mass., 1971.

Globe, C. G. and Desantis, P. V., Optimal design of mixed steel composite girders, *Proc. ASCE, J. Struct. Div.*, **92**, ST 6, 25–34, 1966.

Hadley, G., Linear Programming, Addison-Wesley, Reading, Mass., 1962.

Hadley, G., Nonlinear and Dynamic Programming, Addison-Wesley, Reading, Mass., 1964.

Mangasarian, O. L., Nonlinear Programming, McGraw-Hill, New York, 1969.

Mitra, G. and Wolfenden, K., A computer technique for optimizing the sites and heights of transmission line towers—a dynamic programming approach, *Computer J.*, **10**, 4, 347–51, 1968.

Palmer, A. C., Optimal structural design by dynamic programming, *Proc. ASCE, J. Struct. Div.*, **94**, ST 8, 1887–906, 1968.

Ranyard, J. C. and Wren, A., The optimum arrangement of towers in an electric power transmission line, *Computer J.*, **10**, 2, 157–61, 1967.

Sheu, C. Y. and Prager, W., Recent developments in optimal structural design, *Appl. Mech. Rev.*, **21**, 10, 985–92, 1968.

Templeman, A. B., The Use of Geometric Programming Methods for Structural Optimization, AGARD, Lecture Series, No. 70, Neuilly Sur Seine, France.

PART II STRUCTURAL ANALYSIS

2
Buckling of Columns

2.1 INTRODUCTION

Structural design normally involves the determination of stresses which is based on the assumption that a stable equilibrium exists between internal forces and external loading. Further, it may be noted that any slight change in the loading does not produce excessive stresses.

A structure may have two kinds of failure, namely, (a) material failure and (b) form failure. In material failure, the stresses in the structure exceed the specified safe limit, resulting in the formation of cracks which cause failure. In form failure, though the stresses may not exceed the safe value, they may not be able to maintain the structural forms for which they are designed. Here, the structure does not fail physically but deforms to some other shape due to intolerable external disturbance.

Form failure does not occur in all structures under all loadings. It takes place in structures when the conditions of loading are such that compressive stresses get introduced. To understand the cause of this failure, we need to know not only the equilibrium of the structure but also the nature of the equilibrium. When the magnitude of the load on a structure is such that the equilibrium is changed from stable to neutral, the load is called a *critical load* (also sometimes referred to as a *buckling load*). In other words, the load at which the straight configuration of the column ceases to be stable is the load at which neutral equilibrium is possible. This phenomenon of change of equilibrium is known as the *buckling of structures*.

Even a simple structural element under an axial compressive loading behaves in a complex manner. In structural design, all types of structural sections are used. These can be broadly classified into closed sections and open sections. The mode of form failure of structural elements of these sections depends on (a) flexural rigidity *EI*, (b) torsional rigidity *GJ*, (c) position of shear centre with respect to the centre of gravity, and (d) symmetry of cross-section. In general, for open sections, *GJ* is very small compared with *EI*. In order to study the behaviour of columns and to investigate the effect of each of the four parameters, it is advisable to begin with a simple model and gradually relax the constraints one at a time. In so doing, the questions that arise are: Why does the column buckle? And how is the load to be estimated? These questions can be easily answered if we ascertain whether there is any equilibrium configuration other than the straight one. Such a probe could be conducted in several ways (depending on the method of generating the alternative configurations). We shall confine our study to the most common of these, namely, the equilibrium and energy approaches.

2.2 EQUILIBRIUM APPROACH

The equilibrium approach deals with the equilibrium configuration of the perfect system. Let us consider an idealized column, namely, the Euler column, and derive the equilibrium equation by assuming that
      (i) the member is axially loaded;

(ii) the area of cross-section is uniform throughout the length of the column;
(iii) the material is linearly elastic;
(iv) the deflections are small $(v \ll h)$;
(v) the ends are hinged.

Figure 2.1 shows the straight and deformed configurations of the column. In view of assump-

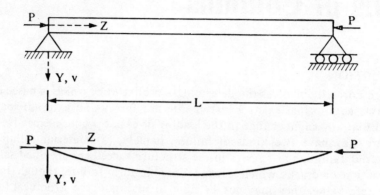

Fig. 2.1   Column under Axial Load.

tion (iv), the internal resistive moment is

$$M = -EI \frac{d^2v}{dz^2},\tag{2.1}$$

where M is the bending moment and I the second moment of area. Equating this to the external moment, we get

$$d^2v/dz^2 + k^2v = 0,\tag{2.2}$$

where $k^2 = P/EI$. The solution of Eq. (2.2), after substituting the boundary conditions for the simply-supported column (Fig. 2.1), is given by

$$v = A \sin \frac{n\pi z}{L}\tag{2.3}$$

and the critical load is

$$P_{cr} = n^2\pi^2 \frac{EI}{L^2}.\tag{2.4}$$

The shape of deformation of the column is given by Eq. (2.3). However, the amplitude is indeterminate and can have any value. From design considerations, critical stress becomes important and is defined as

$$\sigma_{cr} = \frac{n^2\pi^2 E}{(L/\rho)^2},\tag{2.5}$$

where $\rho^2 = I/A_c$. It should be noted that (2.2) cannot be directly used for any other boundary condition.

2.3   DERIVATION OF GENERAL EQUILIBRIUM EQUATION

Consider a column subjected to an axial load P, as shown in Fig. 2.2a. Isolating a small ele-

(a) Column under axial load

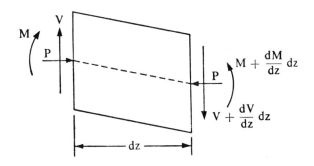

(b) Forces acting on element

Fig. 2.2   Column with Internal Forces.

ment dz and indicating all possible forces and moments, we have, from the equilibrium of transverse forces (see Fig. 2.2b),

$$V + \frac{dV}{dz} dz - V = 0.$$

Hence,

$$\frac{dV}{dz} = 0. \tag{2.6}$$

From the moment equilibrium, we get

$$M - (M + \frac{dM}{dz} dz) + V dz + \frac{dV}{dz} dz \, dz + P \, dv = 0,$$

$$-\frac{dM}{dz} + V + P \frac{dv}{dz} = 0. \tag{2.7}$$

Eliminating V from Eq. (2.7) and using Eq. (2.1), we obtain

$$\frac{d^2}{dz^2} (EI \frac{d^2v}{dz^2}) + P \frac{d^2v}{dz^2} = 0. \tag{2.8}$$

Equation (2.8) is the governing equation for a nonuniform column under axial load. For such a column, it is rather difficult, if not impossible, to obtain a closed-form solution. For columns of uniform cross-section, the use of a nondimensional parameter $\xi (= z/L)$ reduces (2.8) to

$$\frac{d^4v}{d\xi^4} + \lambda^2 \frac{d^2v}{d\xi^2} = 0, \tag{2.9}$$

where $\lambda^2 = PL^2/EI$ is the buckling parameter. The most general solution of Eq. (2.9) is

$$v = C_1 \cos \lambda\xi + C_2 \sin \lambda\xi + C_3\xi + C_4, \tag{2.10}$$

where the constants C_1, C_2, C_3, and C_4 are determined from the boundary conditions of the column.

2.3.1 BOUNDARY CONDITIONS
In a column,

    (i) if both ends are clamped, then

$$v = 0 = \frac{1}{L}\frac{dv}{d\xi} \qquad (\xi = 0, 1); \tag{2.11}$$

    (ii) if one end is clamped and the other end is free, then

$$v = 0 = \frac{1}{L}\frac{dv}{d\xi} \qquad (\xi = 0), \tag{2.12a}$$

$$-\frac{EI}{L^2}\frac{d^2v}{d\xi^2} = 0 \qquad (\xi = 1), \tag{2.12b}$$

$$\frac{d^3v}{d\xi^3} + \lambda^2 \frac{dv}{d\xi} = 0 \qquad (\xi = 1); \tag{2.12c}$$

    (iii) if one end is clamped and the other end is pinned, then

$$v = \frac{1}{L}\frac{dv}{d\xi} = 0 \qquad (\xi = 0),$$

$$v = \frac{-EI}{L^2}\frac{d^2v}{d\xi^2} = 0 \qquad (\xi = 1); \tag{2.13}$$

    (iv) if both ends are pinned, then

$$v = \frac{-EI}{L^2}\frac{d^2v}{d\xi^2} = 0 \qquad (\xi = 0, 1). \tag{2.14}$$

Table 2.1 gives the values of the buckling parameter λ for various boundary conditions for a uniform column.

Table 2.1   Values of Buckling Parameter λ for Uniform Column

End conditions	λ_1	λ_2	λ_3
Both ends simply-supported	π	2π	3π
One end fixed, the other end hinged	4.473	10.903	17.221
One end fixed, the other end free	$\pi/2$	$3\pi/2$	$5\pi/2$
Both ends clamped	2π	4π	6π

Our analysis is thus valid only for ideal boundary conditions which are rare. In practice, the columns are usually connected to one or more members which interact with the deformation of the columns, resulting in the dependence of the boundary reactions on the relative rigidities of the members. This type of support is generally referred to as *elastic restraint*.

Consider a column (see Fig. 2.3) hinged at the lower end and elastically restrained by a beam at the upper end. Here, the boundary conditions are

$$v = -EI\frac{d^2v}{d\xi^2} = 0 \qquad (\xi = 0),$$

$$\left.\frac{dv}{d\xi}\right|_{\xi=1} \text{(for column)} = \psi\left.\right|_{\text{slope of beam}},$$

$$v = 0 \qquad (\xi = 1).$$

(2.15)

The governing equation is given by (2.9) and the general solution by Eq. (2.10). The slope of

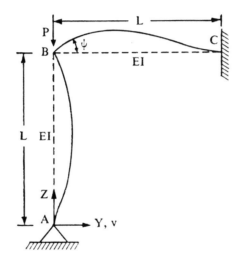

Fig. 2.3   Column with Elastic Support.

the beam at point B is obtained by treating BC as a beam which is fixed at one end and hinged at the other end and subjected to a bending moment M. Then the bending slope is

$$\psi = ML^2/(4EI).$$

(2.16)

From the column study, we get

$$M = -EI\frac{d^2v}{d\xi^2} \qquad (\xi = 1).$$

(2.17)

Therefore,

$$\psi = -\tfrac{1}{4}(\frac{d^2v}{d\xi^2}).$$

(2.18)

Using Eqs. (2.15) and (2.18) together with (2.10), we get the characteristic equation for λ as

$$\tan \lambda = 4\lambda/(4 + \lambda^2).$$

(2.19)

Here, the smallest root of practical interest is $\lambda = 3.83$. Hence,

$$P_{cr} = (14.7)\frac{EI}{L^2}.$$ (2.20)

It may be noted that the value of P_{cr} lies between the values for simply-supported and fixed-end columns.

2.4 ENERGY APPROACH

The energy approach is based on an extremum principle in mechanics which employs an energy criterion and is characterized by the condition of equilibrium in an elastic system. In instances where an exact solution is not possible or is difficult, the energy approach can be used to obtain a good approximate solution, and has the inherent ability to ultimately lead to the exact solution if a large number of terms are taken to represent the deflection shape.

A conservative system is in equilibrium when the energy stored is equal to the work done by the external loads. The equilibrium state before P approaches P_{cr} may be expressed mathematically as

$$U - W_d > 0,$$ (2.21)

where U is the strain energy and W_d the work done by the external load P. As P increases, $U - W_d$ decreases and tends to

$$U - W_d = 0.$$ (2.22)

This indicates the onset of buckling.

The strain energy due to bending is

$$U = \tfrac{1}{2}\int_0^L EI(\frac{d^2v}{dz^2})^2\ dz.$$ (2.23)

The work done by the load P (see Fig. 2.2) in moving through a distance ΔL is

$$W_d = P\Delta L = \frac{P}{2}\int_0^L (\frac{dv}{dz})^2\ dz.$$ (2.24)

From Eq. (2.22), we have

$$P_{cr} = \frac{\displaystyle\int_0^L EI(\frac{d^2v}{dz^2})^2\ dz}{\displaystyle\int_0^L (\frac{dv}{dz})^2\ dz}$$ (2.25)

The critical load P_{cr} can be obtained only if we can assume a displacement function $v(z)$ and its accuracy depends on the choice of this function. The energy approach gives an upper bound to the exact solution. Sometimes, even a one-term approximation satisfying both the geometric and natural boundary conditions provides a good result. When a simple function does not converge to the desired accuracy, a large number of displacement functions which individually satisfy the prescribed geometric boundary conditions may be chosen. Then the principle of stationary potential energy, if approximately applied, gives

$$\delta\pi_p = 0,$$ (2.26)
$$\delta(U + V) = 0.$$ (2.27)

In other words, the total potential energy π_p does not change when the structure passes from its equilibrium configuration to the adjacent configuration.

Let the displacement function be expressed in the form

$$v(z) = a_1\phi_1(z) + a_2\phi_2(z) + a_3\phi_3(z) + \ldots + a_n\phi_n(z), \tag{2.28}$$

where $\phi_1, \phi_2, \ldots$ satisfy the prescribed geometric boundary conditions. The coefficients a_i are obtained from

$$\frac{\partial}{\partial a_i}(U + V) = 0 \qquad (i = 1, 2, \ldots, n). \tag{2.29}$$

This leads to n homogeneous, algebraic, simultaneous equations which, on simplification, result in the characteristic equation. This gives the critical load.

Example 2.1   Consider a uniform column (see Fig. 2.4). Here, the geometric boundary

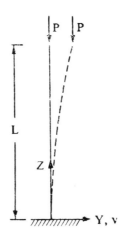

Fig. 2.4   Clamped-Free Column.

conditions are

$$v = 0 \qquad (\xi = 0), \tag{2.30}$$

$$\frac{1}{L}\frac{dv}{d\xi} = 0 \qquad (\xi = 0).$$

A function of the type

$$v = A\xi^2 + B\xi^3 \tag{2.31}$$

satisfies Eqs. (2.30). In terms of the nondimensional parameter ξ, the strain energy due to bending is

$$U = \frac{EI}{2L^3}\int_0^L \left(\frac{d^2v}{d\xi^2}\right)^2 d\xi. \tag{2.32}$$

Substituting for v in Eq. (2.32) and integrating, we get

$$U = \frac{EI}{L^3}(2A^2 + B^2 + 6AB). \tag{2.33}$$

Similarly,

$$V = -\frac{P}{30L}(20A^2 + 45AB + 27B^2), \tag{2.34}$$

$$\delta(U + V) = \frac{\partial}{\partial A}(U + V)\,\delta A + \frac{\partial}{\partial B}(U + V)\,\delta B = 0. \tag{2.35}$$

Since δA and δB are arbitrary, Eq. (2.35) can be satisfied if and only if

$$\frac{\partial}{\partial A}(U + V) = 0, \qquad \frac{\partial}{\partial B}(U + V) = 0. \tag{2.36}$$

Substituting for U and V from Eqs. (2.33) and (2.34) in Eqs. (2.36) and differentiating, we have

$$A(4 - 1.33\lambda^2) + B(6\lambda^2 - 1.5) = 0,$$
$$A(6\lambda^2 - 1.5) + B(12 - 1.8\lambda^2) = 0. \tag{2.37}$$

This leads to the characteristic equation

$$3\lambda^2 - 104\lambda + 240 = 0. \tag{2.38}$$

The smallest root of Eq. (2.38) is $\lambda = 2.49$; hence,

$$P_{cr} = (2.49)\frac{EI}{L^2}. \tag{2.39}$$

This solution differs from the exact solution (see Table 2.1) by slightly less than 1 per cent. However, in all problems, the convergence may not be so fast since the rate of convergence varies from one problem to another. .

2.5 BUCKLING OF OPEN SECTIONS

In Sections 2.1–2.4, we assumed that the bending rigidity EI is much smaller than the torsional rigidity GJ. If this is so, there is a possibility of failure under flexure. Closed and solid sections are prone to this type of failure. Further, the columns of these sections buckle in the plane of principal axes and do not undergo rotation. Such behaviour can be reasonably assumed for any section having two axes of symmetry, but this is not true when only one of the principal axes is an axis of symmetry or when there is no symmetry. When open sections such as channel section, hat section, and angle section are subjected to an axial compressive load, they simultaneously bend and twist. Such a form failure should be investigated since, due to low torsional rigidity, the critical buckling load of the column may be less than the critical load predicted by the flexure formula. Moreover, the critical load for the coupled mode lies between the critical load for the pure torsional mode and that for the pure bending mode. A pure torsional buckling mode exists in sections whose centre of gravity coincides with the shear centre. For any general section, a combined torsional-flexure mode occurs.

2.5.1 TORSIONAL LOAD-DEFORMATION CHARACTERISTICS

It is well known that when a torque is applied to a member whose cross-section is noncircular, the cross-section twists and warps. If warping is freely allowed, then the applied torque is resisted totally by the St. Venant shearing stress. On the other hand, if the member is restrained from warping, then the torque is resisted partly by the St. Venant shearing stress and partly by the stresses produced due to the constraint on warping. The total torque T may be written as

$$T = T_J + T_r,$$ (2.40)

where T_J is the St. Venant torque and T_r the torsion-bending torque.

The distribution of the torque T for an *I*-section which is prevented from warping at one end and is free at the other end is schematically represented in Fig. 2.5.

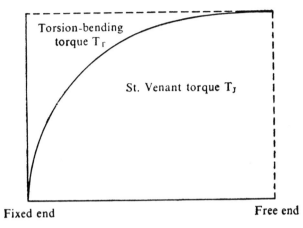

Fixed end   Free end

Fig. 2.5   Variation of Torque with Axis.

Let us consider a bar of constant circular section subjected to a torque T. Here, the torque is resisted completely by the circumferential shear stresses whose magnitude varies as their distance from the centre of the section. It is defined as

$$T = GJ \frac{d\beta}{dz},$$ (2.41)

where J is the St. Venant torsional constant, β the total twist, and z the axial coordinate. The constant J can be obtained by employing the membrane analogy. For open sections, the approximate value of J is given by

$$J = \frac{1}{3} \sum_{i=1}^{n} s_i t_i^3,$$ (2.42)

where s_i is the mean length of the *i*-th element, t_i the thickness of the *i*-th element, and n the number of elements. For different cross-sections, the expressions for J are given in Table 2.2. For hollow sections, J is given by

$$J = \frac{4A^2}{\oint (ds_i/t_i)},$$ (2.43)

where A is the area enclosed by the mean line, s_i the circumferential length, and t_i the thickness of the *i*-th element.

For bars of noncircular sections (e.g., the *I*-section), if warping is freely allowed, no axial stresses develop. However, for constrained sections, the axial stresses vary from zero at the free end of the member to a maximum at the fixed end. The shear forces in the flange member form a couple which is referred to as the torsion-bending torque.

To derive an expression for T_r, let us consider a beam of I-section of length L, subjected to a torque T at the free end (see Fig. 2.6) If the top flange is an independent cantilever beam

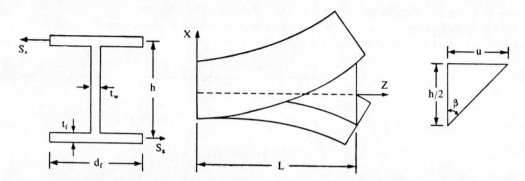

Fig. 2.6   Twisting of I-Section Restrained against Warping.

of width t_f and depth d_f, subjected to a shear load S_x, the torsion-bending torque T_r is

$$T_r = S_x \cdot h, \tag{2.44}$$

where

$$S_x = dM/dz. \tag{2.45}$$

Defining u as the displacement along the X-direction, we get

$$M = -EI_f \frac{d^2u}{dz^2}, \tag{2.46}$$

where I_f is the second moment of area of the flange about the X-axis. For small deformation, u is related to β as

$$u = \frac{h}{2} \beta. \tag{2.47}$$

Therefore,

$$M = -EI_f \frac{h}{2} \frac{d^2\beta}{dz^2}. \tag{2.48}$$

Then T_r may be expressed as

$$T_r = -EI_f \frac{h^2}{2} \frac{d^2\beta}{dz^2}. \tag{2.49}$$

Hence,

$$T = GJ \frac{d\beta}{dz} - E\Gamma \frac{d^3\beta}{dz^3}, \tag{2.50}$$

where

$$\Gamma = I_f \frac{h^2}{2} \quad \text{(for I-section).}$$

Table 2.2 gives the expressions for Γ for some typical sections.

Table 2.2   Values of J and Γ for Some Typical Structural Sections

Cross-section	Shear centre	J	Γ
I-section 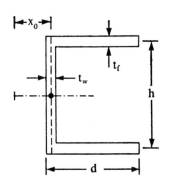	O	$2J_1 + J_2$	$\frac{1}{24}t_f h^2 d^3$ (about O)
Channel section	$y_0 = 0$, $x_0 = d\dfrac{3\rho}{1 + 6\rho}$, where $\rho = d/h$	$2J_1 + J_2$	$\dfrac{t_f h^2 d^3}{12}\dfrac{2 + 3\rho}{1 + 6\rho}$ [about (x_0, y_0)], where $\rho = d/h$
Z-section 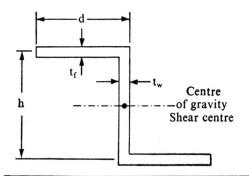	O	$2J_1 + J_2$	$\dfrac{t_f d^3 h^2}{12}\dfrac{2 + \rho}{1 + 2\rho}$ (about O), where $\rho = d/h$

Table 2.2   Values of J and Γ for Some Typical Structural Sections (cont.)

Cross-section	Shear centre	J	Γ
T-section			
	O	$J_1 + J_2$	$\dfrac{t_f^3 h^3}{36} \left(1 + \dfrac{\rho^3}{4}\right)$ (about O), where $\rho = d/h$

$J_1 = (1/3)dt_f^3; \quad J_2 = (1/3)ht_w^3.$

2.6   TORSION-FLEXURE BUCKLING

Consider a general cross-section, as shown in Fig. 2.7. Here, X and Y represent the principal axes of the section passing through its centre of gravity, O stands for the centroid, and C for the shear centre. The total deformation is due to translation and rotation. We shall assume that the cross-section **does** not distort during buckling. The position of OC after deformation is given by $O''C'$. u and v represent the displacement of the point C along the X- and Y-direction.

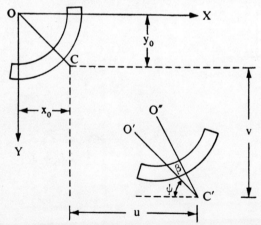

Fig. 2.7   Torsional-Flexural Buckling Deformation.

The deflection of the centroid from the X- and Y-axis can be obtained as follows. Let u' and v' be the displacement of O'' with respect to O. From geometry,

$$u' = x_0 + u - O''C' \cos(\beta + \psi) = x_0 + u - OC \cos\psi + OC\,\beta \sin\psi$$
$$= x_0 + u - x_0 + y_0\beta, \tag{2.51}$$

$$u' = u + y_0\beta. \tag{2.52}$$

Similarly, we can show that

$$v' = v - x_0\beta. \tag{2.53}$$

The differential equations for the deflection curve are

$$\frac{d^2}{dz^2}(EI_y \frac{d^2u}{dz^2}) = -P(\frac{d^2u}{dz^2} + y_0 \frac{d^2\beta}{dz^2}), \tag{2.54}$$

$$\frac{d^2}{dz^2}(EI_x \frac{d^2v}{dz^2}) = -P(\frac{d^2v}{dz^2} - x_0 \frac{d^2\beta}{dz^2}). \tag{2.55}$$

The differential equation for the torsional curve is

$$E\Gamma \frac{d^4\beta}{dz^4} - (GJ - \frac{I_0 P}{A})\frac{d^2\beta}{dz^2} - Px_0 \frac{d^2v}{dz^2} + Py_0 \frac{d^2u}{dz^2} = 0, \tag{2.56}$$

where $I_0 = I_x + I_y + A(x_0^2 + y_0^2)$. Equations (2.54), (2.55), and (2.56) are coupled differential equations. Depending on the symmetry, we have (a) $x_0 = 0$, (b) $y_0 = 0$, and (c) $x_0 = 0$ and $y_0 = 0$. Coupling along one direction takes place for (a) and (b), and torsional buckling independent of flexure occurs for (c).

2.6.1 BUCKLING OF CHANNEL SECTION

Consider a column of length L whose cross-section is shown in Fig. 2.8. Due to symmetry

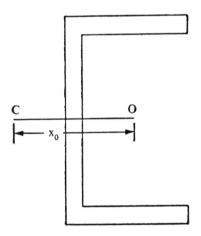

Fig. 2.8 Column with Uniform Channel Section.

about the X-axis, $y_0 = 0$. Substituting this value in (2.54) and (2.56), we have, for a uniform column,

$$EI_y \frac{d^4u}{dz^4} + P\frac{d^2u}{dz^2} = 0, \tag{2.57}$$

$$EI_x \frac{d^4v}{dz^4} + P(\frac{d^2v}{dz^2} - x_0 \frac{d^2\beta}{dz^2}) = 0, \tag{2.58}$$

$$E\Gamma \frac{d^4\beta}{dz^4} - (GJ - \frac{I_0 P}{A})\frac{d^2\beta}{dz^2} - Px_0 \frac{d^2v}{dz^2} = 0. \tag{2.59}$$

It is evident from these equations that the deformations v and β are coupled. Therefore, Eq. (2.57) can be solved independent of Eqs. (2.58) and (2.59). The mode of buckling in the plane of symmetry is one of pure flexure, and that perpendicular to the plane of symmetry is coupled with torsion.

By defining $z = \xi L$, we write (2.58) and (2.59) in the nondimensional form

$$\frac{d^4v}{d\xi^4} + \frac{L^2P}{EI}\left(\frac{d^2v}{d\xi^2} - x_0\frac{d^2\beta}{d\xi^2}\right) = 0, \tag{2.60}$$

$$\frac{E\Gamma}{L^2}\frac{d^4\beta}{d\xi^4} - \left(GJ - \frac{I_0P}{A}\right)\frac{d^2\beta}{d\xi^2} - Px_0\frac{d^2v}{d\xi^2} = 0. \tag{2.61}$$

If we assume that the ends of the column are simply-supported, that is, free to warp about the X-axis but restrained from rotation about the Z-axis, the boundary conditions are

$$v = \beta = 0 \qquad (\xi = 0, 1),$$

$$\frac{d^2v}{d\xi^2} = \frac{d^2\beta}{d\xi^2} = 0^{\cdot} \qquad (\xi = 0, 1). \tag{2.62}$$

Equations (2.60) and (2.61) are linear differential equations of even order, and we can look for a solution of the type

$$v = A \sin \pi\xi, \qquad \beta = B \sin \pi\xi. \tag{2.63}$$

This implies that the changes in the mode shape due to coupling are of the second order and can therefore be ignored. The substitution of solution (2.63) in (2.60) and (2.61) and subsequent simplification yield the characteristic equation

$$P_{cr}^2\left(1 + \frac{x_0^2A}{I_0}\right) - (P_x + P_\beta)P_{cr} + P_xP_\beta = 0, \tag{2.64}$$

where P_x and P_β are the critical buckling loads in flexural and torsional modes, respectively. This quadratic equation gives two solutions for P_{cr}: When one is smaller than P_x, the other is larger than P_β; and when one is smaller than P_β, the other is larger than P_x. The critical load for the column is obtained by choosing the smaller of the two roots or the buckling load in the plane of symmetry, whichever is the minimum.

2.7 NUMERICAL TECHNIQUES

The foregoing techniques are useful if it is possible to get a closed-form solution which can be incorporated in the computer program. But this is rather rare. Often, it is necessary to seek a numerical solution. Of the available techniques, the finite difference and finite element techniques are suitable for computer application. The scope of the former is limited to well-defined boundaries, whereas that of the latter is more general and can therefore be applied to all types of problems. Both the techniques have the ability to lead to the exact solution.

2.7.1 FINITE DIFFERENCE TECHNIQUE

The numerical solution of a differential equation essentially entails defining the numerical values of the unknown integrals at some pivotal points spaced along one coordinate axis for ordinary differential equations and along a plane for two-dimensional partial differential equations. When the pivotal points are evenly-spaced, the Taylor series may be applied in conjunction with the concept of difference. Depending on the definition of the derivatives at a point in terms of the differences of the neighbouring points, we can obtain a large variety of practical approximations for the derivatives. Notable among these are (a) backward differences defined by all the points to the left of the point i; (b) forward differences expressed in terms of all the points to the right of the point i; and (c) central differences in which the pivotal points are symmetrically located (see Fig. 2.9) with respect to the point i. Central differences, because of their reasonable accuracy, are particularly useful in the solution of boundary value problems.

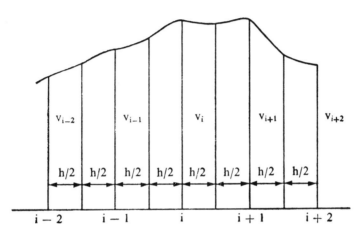

Fig. 2.9 Pivotal Points for Central Differences.

Let the function $v(z)$ be known at evenly-spaced points and also at the mid-points of evenly-spaced points. Then the first central difference of $v(z)$ at i is defined by

$$\Delta v_i = [v(z_i + \tfrac{h}{2}) - v(z_i - \tfrac{h}{2})]/h = (v_{i+1/2} - v_{i-1/2})/h. \tag{2.65}$$

The second difference at i is described by

$$\Delta^2 v_i \equiv \Delta(\Delta v_i) = [\{v_{(i+1/2)+1/2} - v_{(i-1/2)+1/2}\} - \{v_{(i+1/2)-1/2} - v_{(i-1/2)-1/2}\}]/h^2$$
$$= (v_{i+1} - 2v_i - v_{i-1})/h^2. \tag{2.66}$$

Similarly, the third and fourth differences may be written as

$$\Delta^3 v_i = (v_{i+3/2} - 3v_{i+1/2} + 3v_{i-1/2} - v_{i-3/2})/h^3, \tag{2.67}$$

$$\Delta^4 v_i = (v_{i+2} - 4v_{i+1} + 6v_i - 4v_{i-1} + v_{i-2})/h^4. \tag{2.68}$$

Let us consider the flexure buckling of a uniform column fixed at one end and hinged at the other end, as shown in Fig. 2.10. In this case, the governing equation is given by (2.9) and the boundary conditions are

$$v(0) = \frac{1}{L}\frac{dv}{d\xi}(0) = v(1) = \frac{1}{L}\frac{d^2v}{d\xi^2}(1) = 0. \tag{2.69}$$

Dividing the interval $(0, 1)$ of ξ into n equal parts of length $h = 1/n$, and substituting Eqs. (2.66) and (2.68) in (2.9), we get the finite difference equation

$$v_{i+2} - 4v_{i+1} + 6v_i - 4v_{i-1} + v_{i-2} + \lambda_n(v_{i+1} - 2v_i + v_{i-1}) = 0, \tag{2.70}$$

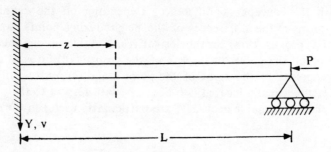

Fig. 2.10   Clamped-Hinged Column under Axial Load.

where

$$\lambda_n = \frac{PL^2}{n^2 EI}.$$

(2.71)

Equation (2.70) may be rewritten as

$$v_{i+2} + (\lambda_n - 4)v_{i+1} + (6 - 2\lambda_n)v_i + (\lambda_n - 4)v_{i-1} + v_{i-2} = 0.$$

(2.72)

This holds at $(n-1)$ internal points, namely, $i = 1, 2, \ldots, n-1$. The boundary conditions (2.69) may be expressed also in terms of central differences which are applied at the end points, namely, $i = 0$ and $i = n$, as

$$v_0 = 0, \qquad v_{-1} = v_1, \qquad v_n = 0, \qquad v_{n+1} = -v_{n-1}.$$

(2.73)

Equation (2.72) leads to a set of $(n-1)$ linear, algebraic, homogeneous equations, and has a nontrivial solution only if the determinant of the coefficients of the v_i's is identically zero. It is always advisable to start the solution with a small value of n and increase this in stages. This results in a saving of both computation time and storage. Also, the convergence of the solution can be checked. For $n = 2$, the characteristic equation reduces to

$$(6 - 2\lambda_2)v_1 = 0,$$

(2.74)

$$\lambda_2 = 3,$$

$$P_{cr} = 12 \frac{EI}{L^2}.$$

(2.75)

For $n = 3$, the two linear homogeneous equations are

$$(7 - 2\lambda_3)v_1 + (\lambda_3 - 4)v_2 = 0, \qquad (\lambda_3 - 4)v_1 + (5 - 2\lambda_3)v_2 = 0,$$

(2.76)

and hence the characteristic equation is

$$3\lambda_3^2 - 16\lambda_3 + 19 = 0,$$

(2.77)

whose smallest root is 1.785, and

$$P_{cr} = (16.063) \frac{EI}{L^2}.$$

(2.78)

For $n = 4$, it can be shown that

$$P_{cr} = (17.772) \frac{EI}{L^2},$$

(2.79)

whereas the exact value is

$$P_{cr} = (20.187) \frac{EI}{L^2} \tag{2.80}$$

Thus, we observe that the accuracy of the solution can be improved by reducing the spacing h between the pivotal points. Once the spacing is reduced, approximate values of even higher buckling loads can be obtained. In addition, any of the standard programs may be used to solve for the roots.

2.7.2   FINITE ELEMENT TECHNIQUE

Most numerical techniques lead to solutions that yield approximate values of unknown quantities only at selected points in a body.

A body can be discretized into an equivalent system of smaller bodies. The assemblage of such systems then represents the original body. Each subsystem is solved first, and then combined with the other subsystems to obtain the solution for the whole system. Of the numerical techniques, the finite element technique is the most suitable for digital computers. It is applicable to a wide range of problems involving nonhomogeneous materials, nonlinear stress-strain relations, and complicated boundary conditions. Such problems are usually tackled by one of three approaches, namely, (a) the displacement (or stiffness) method, (b) the equilibrium (or force) method, and (c) the mixed method. The displacement method (to which we shall confine our discussion) is widely used because of the simplicity with which it can be handled on a computer. Besides, its application to both statically determinate and statically indeterminate problems is the same.

In the stiffness approach, a structure is divided into a number of finite elements which are interconnected at joints termed *nodes*. Simple displacement functions are then chosen to represent the variation of displacement over each element. The amplitudes of these functions are the displacements and/or the derivatives of the displacements at the nodes. A displacement function is generally expressed in terms of the polynomials.

The potential energy (π_p) for an element of a structure may be defined as

$$
\begin{aligned}
(\pi_p)_e = {}& \text{strain energy } (U) + \text{potential energy of applied forces (or negative} \\
& \text{of work done)}
\end{aligned}
$$

$$
\begin{aligned}
= {}& \tfrac{1}{2} \int_{v^{(e)}} e^T \sigma \, dv - \int_{s_1^{(e)}} (\varphi_x u + \varphi_y v + \varphi_z w) \, ds_1 \\
& - \int_{v^{(e)}} (\overline{X}_x u + \overline{Y}_y v + \overline{Z}_z w) \, dv,
\end{aligned} \tag{2.81}
$$

where s_1 is the surface of the body on which the surface forces φ_x, φ_y, and φ_z are imposed; $\overline{X}_x$, $\overline{Y}_y$, and $\overline{Z}_z$ are prescribed body forces per unit volume; u, v, and w are displacements along the X-, Y-, and Z-direction; and σ and e represent all the six components of stress and strain for a three-dimensional body. Since, in the finite element technique, the body is considered an assemblage of elements, Eq. (2.81), for the entire structure, may be expressed as

$$
\begin{aligned}
\pi_p = {}& \tfrac{1}{2} \sum_{e=1}^{N} \int_{v^{(e)}} e^T \sigma \, dv - \sum_{e=1}^{N} \int_{s_1^{(e)}} (\varphi_x u + \varphi_y v + \varphi_z w) \, ds_1 \\
& - \sum_{e=1}^{N} \int_{v^{(e)}} (\overline{X}_x u + \overline{Y}_y v + \overline{Z}_z w) \, dv,
\end{aligned} \tag{2.82}
$$

where N is the total number of elements of the structure.

For a one-dimensional structure, the displacement $u(x)$ may be written in the polynomial form

$$u(x) = \alpha_1 + \alpha_2 x + \alpha_3 x^2 + \ldots + \alpha_{n+1} x^n, \tag{2.83}$$

where $\alpha_1, \alpha_2, \ldots, \alpha_{n+1}$ are called the *generalized coordinates* or *generalized displacement amplitudes*. The higher the value of n, the greater is the accuracy with which the displacement amplitudes can be represented. These are called generalized because they need not represent the physical displacement of the element on a one-to-one basis, but are usually linear combinations of some of the nodal displacements and the derivatives of these displacements. In matrix notation, Eq. (2.83) is given by

$$u(x) = \phi^T \alpha, \tag{2.84}$$

where

$$\begin{aligned} \phi^T &= [1, x, x^2, \ldots, x^n], \\ \alpha^T &= \{\alpha_1, \alpha_2, \ldots, \alpha_{n+1}\}. \end{aligned} \tag{2.85}$$

For a three-dimensional body, the relation between the nodal degrees of freedom q_1, $q_2, \ldots, q_n$ and the generalized coordinates $\alpha_1, \alpha_2, \ldots, \alpha_n$ may be expressed as

$$\{u\} = [\phi]\{\alpha\}. \tag{2.86}$$

If $q_1 = u$ (at node 1) and $q_2 = u$ (at node 2), then $q_M = u$ (at node M). Equation (2.86) gives

$$\{q\} = \begin{Bmatrix} u_{\text{node 1}} \\ u_{\text{node 2}} \\ \vdots \\ u_{\text{node } M} \end{Bmatrix} = \begin{bmatrix} [\phi]_{\text{node 1}} \\ [\phi]_{\text{node 2}} \\ \vdots \\ [\phi]_{\text{node } M} \end{bmatrix} \{\alpha\},$$

$$\{q\} = [A]\{\alpha\}, \tag{2.87}$$

where M is the total number of nodes for the element being considered. Substituting for $\{\alpha\}$ from Eq. (2.87) in (2.86), we have

$$\{u\} = [\phi][A]^{-1}\{q\} = [N]\{q\}. \tag{2.88}$$

Since the strains are derivatives of the displacements,

$$\{e\} = [B]\{q\}, \tag{2.89}$$

where the matrix $[B]$ is obtained from the derivative of the matrix $[N]$. The strains can also be written as

$$\{e\} = [B_\alpha]\{\alpha\}. \tag{2.90}$$

The matrix $[B]$ is related to $[B_\alpha]$ as

$$[B] = [B_\alpha][A]^{-1}. \tag{2.91}$$

If the stress-strain relation is written in a matrix form as

$$\{\sigma\} = [\chi]\{e\}, \tag{2.92}$$

where $[\chi]$ is the matrix of material constant, then, on substitution for $\{e\}$, we get

$$\{\sigma\} = [\chi][B_\alpha]\{\alpha\}, \tag{2.93}$$

$$\{\boldsymbol{\sigma}\} = [\chi][B]\{\boldsymbol{q}\}. \tag{2.94}$$

Substituting Eqs. (2.88), (2.93), and (2.94) in (2.81), we obtain

$$(\pi_{\mathrm{p}})_{\mathrm{e}} = \tfrac{1}{2} \int_{\mathrm{v}^{(\mathrm{e})}} \boldsymbol{q}^T [B]^T [\chi][B] \boldsymbol{q} \ dv - \int_{s_1^{(\mathrm{e})}} \boldsymbol{q}^T [N]^T \boldsymbol{\varphi} \ ds_1 - \int_{\mathrm{v}^{(\mathrm{e})}} \boldsymbol{q}^T [N]^T X \ dv. \tag{2 95}$$

Applying the variational principle $\delta \pi_{\mathrm{p}} = 0$, we get

$$\delta \boldsymbol{q}^T \left(\int_{\mathrm{v}^{(\mathrm{e})}} [B^T][\chi][B] \ dv \ \boldsymbol{q} - \int_{\mathrm{v}^{(\mathrm{e})}} [N]^T X \ dv - \int_{s_1^{(\mathrm{e})}} [N]^T \boldsymbol{\varphi} \ ds_1 \right) = 0. \tag{2.96}$$

Since $\delta \boldsymbol{q}^T$ is arbitrary, the expression within the parentheses must vanish in order that Eq. (2.96) is satisfied. This gives the equilibrium equation for an element as

$$[k] \boldsymbol{q}^{(\mathrm{e})} = \boldsymbol{Q}^{(\mathrm{e})}, \tag{2.97}$$

where $[k]$ is the elemental stiffness matrix given by

$$[k] = \int_{\mathrm{v}^{(\mathrm{e})}} [B]^T [\chi][B] \ dv \tag{2.98}$$

and $\boldsymbol{Q}^{(\mathrm{e})}$ is the external load vector defined by

$$Q^{(\mathrm{e})} = \int_{\mathrm{v}^{(\mathrm{e})}} [N]^T X \ dv + \int_{s_1^{(\mathrm{e})}} [N]^T \boldsymbol{\varphi} \ ds_1. \tag{2.99}$$

As stated in Section 2.4, the linear stability problem is formulated by applying the principle of stationary total potential energy

$$\delta(U - W_{\mathrm{d}}) = 0. \tag{2.100}$$

In terms of finite elements, U and W_{d} may be written as

$$U = \tfrac{1}{2} \{\boldsymbol{q}\}^T [K]\{\boldsymbol{q}\}, \tag{2.101}$$

$$W_{\mathrm{d}} = \frac{\lambda}{2} \{\boldsymbol{q}\}^T [G]\{\boldsymbol{q}\}, \tag{2.102}$$

where $[K]$ and $[G]$ are the elastic stiffness matrix and the geometric stiffness matrix, respectively, for the total structure, $\{\boldsymbol{q}\}$ the vector of global displacements, and λ the eigenvalue.

Equation (2.98) gives the elastic stiffness matrix for an element. The total stiffness matrix can be obtained by adding the element stiffness matrices.

To get $[G]$, let us consider an element subjected to the inplane stresses σ_x, σ_y, and τ_{xy}. The work done by these stresses due to lateral deformation of the element is given by

$$W_{\mathrm{d}} = \tfrac{1}{2} \sigma_x \int_{\mathrm{v}^{(\mathrm{e})}} \left(\frac{\partial u}{\partial x}\right)^T \left(\frac{\partial u}{\partial x}\right) \ dv + \tfrac{1}{2} \sigma_y \int_{\mathrm{v}^{(\mathrm{e})}} \left(\frac{\partial u}{\partial y}\right)^T \left(\frac{\partial u}{\partial y}\right) \ dv + \tau_{xy} \int_{\mathrm{v}^{(\mathrm{e})}} \left(\frac{\partial u}{\partial x}\right)^T \left(\frac{\partial u}{\partial y}\right) \ dv. \tag{2.103}$$

Using (2.86), i.e., $\{\boldsymbol{u}\} = [\phi]\{\boldsymbol{\alpha}\}$, we get

$$W_{\mathrm{d}} = \tfrac{1}{2} \sigma_x \{\boldsymbol{\alpha}\}^T \left[\int_{\mathrm{v}^{(\mathrm{e})}} \left(\frac{\partial \phi}{\partial x}\right)^T \left(\frac{\partial \phi}{\partial x}\right) \ dv \right] \{\boldsymbol{\alpha}\} + \tfrac{1}{2} \sigma_y \{\boldsymbol{\alpha}\}^T \left[\int_{\mathrm{v}^{(\mathrm{e})}} \left(\frac{\partial \phi}{\partial y}\right)^T \left(\frac{\partial \phi}{\partial y}\right) \ dv \right] \{\boldsymbol{\alpha}\}$$
$$+ \tau_{xy} \{\boldsymbol{\alpha}\}^T \left[\int_{\mathrm{v}^{(\mathrm{e})}} \left(\frac{\partial \phi}{\partial x}\right)^T \left(\frac{\partial \phi}{\partial y}\right) \ dv \right] \{\boldsymbol{\alpha}\} \tag{2.104}$$

Taking the variation with respect to $\boldsymbol{\alpha}$, we obtain

$$\{\delta \boldsymbol{\alpha}\}^T \left[\sigma_x \int_{\mathrm{v}^{(\mathrm{e})}} \left(\frac{\partial \phi}{\partial x}\right)^T \left(\frac{\partial \phi}{\partial x}\right) \ dv + \sigma_y \int_{\mathrm{v}^{(\mathrm{e})}} \left(\frac{\partial \phi}{\partial y}\right)^T \left(\frac{\partial \phi}{\partial y}\right) \ dv + 2\tau_{xy} \int_{\mathrm{v}^{(\mathrm{e})}} \left(\frac{\partial \phi}{\partial x}\right)^T \left(\frac{\partial \phi}{\partial y}\right) \ dv \right] \tag{2.105}$$

In terms of generalized coordinates, the geometric stiffness matrix $[g_\alpha]$ is given by

$$[g_\alpha] = \sigma_x \int_{v^{(e)}} (\frac{\partial \phi}{\partial x})^T (\frac{\partial \phi}{\partial x}) \, dv + \sigma_y \int_{v^{(e)}} (\frac{\partial \phi}{\partial y})^T (\frac{\partial \phi}{\partial y}) \, dv + 2\tau_{xy} \int_{v^{(e)}} (\frac{\partial \phi}{\partial x})^T (\frac{\partial \phi}{\partial y}) \, dv.$$

By using (2.87), we obtain the geometric stiffness matrix $]g]$, in terms of the nodal displacements $\{q\}$, as

$$[g] = [A^{-1}]^T [g_\alpha][A]. \tag{2.106}$$

If we know $[g]$ for an element, we can get the total geometric stiffness matrix $[G]$. There are two methods for obtaining $[g]$: (a) the consistent geometric stiffness technique and (b) the nonconsistent geometric stiffness technique. In the former, the same displacement function is used to derive both the elastic stiffness matrix and the geometric stiffness matrix, whereas, in the latter, a lower-order displacement function is employed to get the geometric stiffness matrix. The main advantage of the consistent geometric stiffness method is that the mode shapes and the eigenvalues are more accurate and are proven upper bounds to the exact solution. In the nonconsistent stiffness approach, on the other hand, it is not possible to prove *a priori* how the solution will converge.

As an example of the application of the finite element technique, let us consider a uniform column of length L. To evaluate the matrices $[K]$ and $[G]$, suppose the beam element has, at each node, two degrees of freedom, one corresponding to translation and one to rotation. Thus, there are four degrees of freedom for an element; the displacement function is therefore assumed to be

$$v = \alpha_1 + \alpha_2 z + \alpha_3 z^2 + \alpha_4 z^3. \tag{2.107}$$

This satisfies the conditions of constant shear and linearly-varying bending moment existing in the beam element. If v_1, v_2, v_3, and v_4 represent the nodal displacements, the displacement v at any point within the element is given by

$$v = [(-\frac{2z^3}{l^3} + 3\frac{z^2}{l^2} - 1)(z - \frac{2z^2}{l} + \frac{z^3}{l^2})(-\frac{3z^2}{l^2} + \frac{2z^3}{l^3})(\frac{z^3}{l^2} - \frac{z^2}{l})] \begin{Bmatrix} v_1 \\ v_2 \\ v_3 \\ v_4 \end{Bmatrix}. \tag{2.108}$$

Thus,

$$\epsilon_z = -y\frac{\partial^2 v}{\partial z^2} = -y[(-\frac{12z}{l^3} + \frac{6}{l^2})(\frac{6z}{l^2} - \frac{4}{l})(-\frac{6}{l^2} + \frac{12z}{l^3})(\frac{6z}{l^2} - \frac{2}{l})] \begin{Bmatrix} v_1 \\ v_2 \\ v_3 \\ v_4 \end{Bmatrix}, \tag{2.109}$$

$$\{e\} = [B]\{q\},$$

where

$$[B] = -y[(-\frac{12z}{l^3} + \frac{6}{l^2})(\frac{6z}{l^2} - \frac{4}{l})(-\frac{6}{l^2} + \frac{12z}{l^3})(\frac{6z}{l^2} - \frac{2}{l})]. \tag{2.110}$$

From (2.98), we obtain

$$[k] = \int_{v^{(e)}} [B]^T [\chi][B] \, dA \, dz = EI \int_0^l [B]^T [B] \, dz. \tag{2.111}$$

By substituting the value of $[B]$ in Eq. (2.111) and integrating, we get

$$[k] = EI \begin{array}{c} \begin{array}{cccc} v_1 & v_2 & v_3 & v_4 \end{array} \\ \begin{bmatrix} \dfrac{12}{l^3} & -\dfrac{6}{l^2} & -\dfrac{12}{l^3} & -\dfrac{6}{l^2} \\[2mm] -\dfrac{6}{l^2} & \dfrac{4}{l} & \dfrac{6}{l^2} & \dfrac{2}{l} \\[2mm] -\dfrac{12}{l^3} & \dfrac{6}{l^2} & \dfrac{12}{l^3} & \dfrac{6}{l^2} \\[2mm] -\dfrac{6}{l^2} & \dfrac{2}{l} & \dfrac{6}{l^2} & \dfrac{4}{l} \end{bmatrix} \end{array} \qquad (2.112)$$

Similarly, we can obtain the consistent geometric stiffness matrix $[g]$ as

$$[g] = P \begin{array}{c} \begin{array}{cccc} v_1 & v_2 & v_3 & v_4 \end{array} \\ \begin{bmatrix} \dfrac{6}{5l} & -\dfrac{1}{10} & -\dfrac{6}{5l} & -\dfrac{1}{10} \\[2mm] -\dfrac{1}{10} & \dfrac{2l}{5} & \dfrac{1}{10} & -\dfrac{l}{30} \\[2mm] -\dfrac{6}{5l} & \dfrac{1}{10} & \dfrac{6}{5l} & \dfrac{1}{10} \\[2mm] -\dfrac{1}{10} & -\dfrac{l}{30} & \dfrac{1}{10} & \dfrac{2l}{15} \end{bmatrix} \end{array} \qquad (2.113)$$

Consider the simply-supported column to be divided into two elements and the nodal displacement vector to be $\{v_1, v_2, v_3, v_4, v_5, v_6\}^T$, where v_1, v_3, and v_5 represent displacements and v_2, v_4, and v_6 denote rotations. This is illustrated in Fig. 2.11. Now the elastic stiffness matrix $[K]$

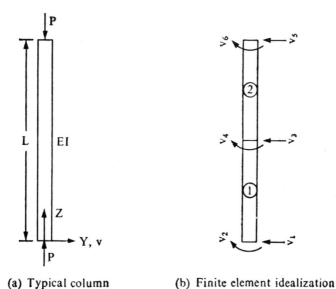

(a) Typical column   (b) Finite element idealization

Fig. 2.11 Uniform Column under Axial Load.

for the structure can be written as

$$[K] = \frac{EI}{l^3}
\begin{matrix}
v_1 & v_2 l & v_3 & v_4 l & v_5 & v_6 l
\end{matrix}
\begin{bmatrix}
12 & -6 & -12 & -6 & 0 & 0 \\
-6 & 4 & 6 & 2 & 0 & 0 \\
-12 & 6 & 24 & 0 & -12 & -6 \\
-6 & 2 & 0 & 8 & 6 & 2 \\
0 & 0 & -12 & 6 & 12 & 6 \\
0 & 0 & -6 & 2 & 6 & 4
\end{bmatrix}
\tag{2.114}$$

and the geometric stiffness matrix $[G]$ as

$$[G] = -\frac{P}{l}
\begin{matrix}
v_1 & v_2 l & v_3 & v_4 l & v_5 & v_6 l
\end{matrix}
\begin{bmatrix}
6/5 & -1/10 & -6/5 & -1/10 & 0 & 0 \\
-1/10 & 2/15 & 1/10 & -1/30 & 0 & 0 \\
-6/5 & 1/10 & 12/5 & 0 & -6/5 & -1/10 \\
-1/10 & -1/30 & 0 & 4/15 & 1/10 & -1/30 \\
0 & 0 & -6/5 & 1/10 & 6/5 & 1/10 \\
0 & 0 & -1/10 & -1/30 & 1/10 & 2/15
\end{bmatrix}.
\tag{2.115}$$

From the boundary conditions of the simply-supported column, we have, for symmetric modes, $v_1 = v_4 = v_5 = 0$. The governing equation for buckling is

$$[[K] - \lambda[G]]\{v\} = 0. \tag{2.116}$$

Since $\{v\}$ is not zero for a nontrivial solution, the determinant $|K - \lambda G| = 0$. That is,

$$\begin{vmatrix}
\dfrac{4EI}{l^3} - \dfrac{2P}{15l} & \dfrac{6EI}{l^3} - \dfrac{P}{10l} \\[2mm]
\dfrac{12EI}{l^3} - \dfrac{Pl}{5} & \dfrac{24EI}{l^3} - \dfrac{12P}{5l}
\end{vmatrix} = 0. \tag{2.117}$$

This leads to a characteristic equation from which the lowest value of the critical load P_{cr} is given by

$$P_{cr} = \frac{(2.45)EI}{l^2} = \frac{(9.8)EI}{L^2}, \tag{2.118}$$

$$P_{cr\,(exact)} = \frac{\pi^2 EI}{L^2}. \tag{2.119}$$

The finite element method can be used also for higher modes and nonuniform columns.

BIBLIOGRAPHY

Bleich, F., Buckling Strength of Metal Structures, McGraw-Hill, New York, 1952.

Burge, C. G. (ed.), Handbook of Aeronautics, Vol. I—Structural principles and data, New Era Publishing Company, London, 1954.

Chajes, A., Principles of Structural Stability Theory, Prentice-Hall, Englewood Cliffs, New Jersey, 1974

Desai, C. S. and Abel, J. F., Introduction to the Finite Element Method, Van Nostrand Reinhold, New York, 1972.

Gerard, G., Introduction to Structural Stability Theory, McGraw-Hill, New York, 1962.

Kuhn, P., Stresses in Aircraft and Shell Structures, McGraw-Hill, New York, 1956.

Przemieniecki, J. S., Theory of Matrix Structural Analysis, McGraw-Hill, New York, 1968.

Salvadori, M. G. and Baron, M. L., Numerical Methods in Engineering, Prentice-Hall, Englewood Cliffs, New Jersey, 1961.

Timoshenko, S. P. and Gere, J. M., Theory of Elastic Stability, McGraw-Hill, New York, 1961.

3

Buckling of Plates

3.1 INTRODUCTION

In Chapter 2, we discussed the buckling of one-dimensional members. In such members, buckling is normally confined to one vertical plane, and the number of boundary conditions is only four. The buckling of plates, however, involves buckling in two planes and the number of boundary conditions is two along each of the four edges. It is therefore a two-dimensional problem, and the deflections, stresses, and strains are functions of two variables. Consequently, the behaviour of plates is more complex.

The basic difference between a column and a plate lies in the buckling characteristics. Once a column buckles, it cannot resist any axial load, and the critical load can be taken as the failure load. On the other hand, a plate can resist any axial load even after the primary buckling load is reached and does not fail until the load is 10–15 times the primary buckling load. In such cases, it is necessary to determine the load-carrying capacity of a plate by considering its postbuckling behaviour. In Chapter 2, we also discussed I- and channel section as well as others, assuming that the web and flange dimensions are very small as compared with the length of the column. However, if they are of the same order, then the flange and web members should be considered as plates with elastic restraint along the edges where web and flange meet. The analysis developed in this chapter is also applicable to flat surfaces such as those of aircraft wings and ship hulls.

3.2 THEORY OF THIN PLATES (Small Deflection Theory)

Figures 3.1a and 3.1b show the coordinate system and the stresses acting on a plate element,

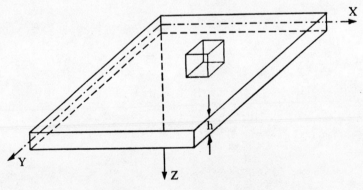

(a) Coordinate system for the plate

Fig. 3.1 Typical Plate Element with Stresses (cont.).

respectively. For thin plates, the effect of shear deformations may be neglected and the stresses considered to be two-dimensional.

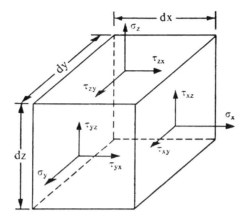

(b) Stresses acting on a plate element

Fig. 3.1   Typical Plate Element with Stresses.

We make two assumptions regarding the behaviour of thin plates:

(i) The shear strains τ_{xz} and τ_{yz} are negligible, and the planes normal to the middle surface before bending remain straight after bending (Kirchhoff's assumption).

(ii) The normal stress σ_z and the corresponding strain ϵ_z are negligible and, therefore, the transverse deflection at any point $(x, y, 0)$ is equal to that at the corresponding point (x, y, z) along the middle surface.

3.3   EQUILIBRIUM APPROACH

To obtain the governing equation for plate buckling, let us consider a plate subjected to the inplane forces N_x, N_y, and N_{xy} (see Fig. 3.2), each defined as force per unit length. The action of

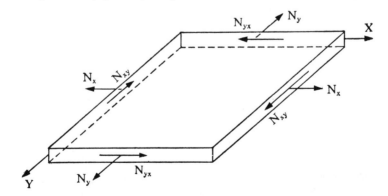

Fig. 3.2   Inplane Forces Acting on Plate Element.

the inplane forces is to bend the plate in the transverse direction. Hence, any element of the plate is in equilibrium under the influence of the external inplane forces and that of the moments and shear forces resulting from the deflection of the plate. In view of the small deflection theory assumption, we may neglect the inplane forces that develop due to bending.

Figure 3.3 shows the deflected shape of the element $dx\, dy$ of the plate in the XY-plane. The sum of the forces along the X- and Y-axis is identically equal to zero. As the plate is

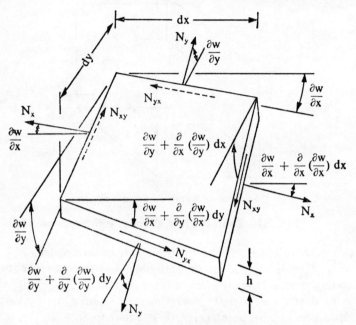

Fig. 3.3   Inplane Forces Acting on Deflected Plate.

deflected, there are components of the inplane forces along the Z-direction. The components of the N_x forces in the Z-direction are

$$N_x\left(\frac{\partial w}{\partial x} + \frac{\partial^2 w}{\partial x^2}\, dx\right) dy - N_x \frac{\partial w}{\partial x}\, dy$$

or

$$N_x \frac{\partial^2 w}{\partial x^2}\, dx\, dy. \tag{3.1}$$

Similarly, we can write the components of N_y and N_{xy} along the Z-direction. The resultant of the inplane forces along the Z-direction is

$$\left(N_x \frac{\partial^2 w}{\partial x^2} + N_y \frac{\partial^2 w}{\partial y^2} + 2N_{xy} \frac{\partial^2 w}{\partial x\, \partial y}\right) dx\, dy. \tag{3.2}$$

In addition to these inplane forces on the slightly bent plate element, there exist shear forces and moments. These are considered positive when acting in the direction indicated in Fig. 3.4. The moments and shear forces per unit length on the negative face of the element are defined as

$$Q_x = \int_{-h/2}^{h/2} \tau_{xz}\, dz, \qquad Q_y = \int_{-h/2}^{h/2} \tau_{yz}\, dz; \tag{3.3a}$$

$$M_x = \int_{-h/2}^{h/2} \sigma_x z \, dz, \qquad M_y = \int_{-h/2}^{h/2} \sigma_y z \, dz; \qquad (3.3b)$$

$$M_{xy} = -\int_{-h/2}^{h/2} \tau_{xy} z \, dz, \qquad M_{yx} = \int_{-h/2}^{h/2} \tau_{yx} z \, dz. \qquad (3.3c)$$

Since the moments and shear forces are functions of x and y, the corresponding quantities along the positive face of the element are

$$Q_x + \frac{\partial Q_x}{\partial x} dx, \qquad M_x + \frac{\partial M_x}{\partial x} dx, \qquad M_{xy} + \frac{\partial M_{xy}}{\partial x} dx,$$

$$Q_y + \frac{\partial Q_y}{\partial y} dy, \qquad M_y + \frac{\partial M_y}{\partial y} dy, \cdot \qquad M_{yx} + \frac{\partial M_{yx}}{\partial y} dy. \qquad (3.4)$$

In the Z-direction, the resultant of the components of the shear forces is

$$\frac{\partial Q_x}{\partial x} dx \, dy + \frac{\partial Q_y}{\partial y} dx \, dy. \qquad (3.5)$$

Thus, the equilibrium equation in the Z-direction is

$$\frac{\partial Q_x}{\partial x} + \frac{\partial Q_y}{\partial y} + N_x \frac{\partial^2 w}{\partial x^2} + N_y \frac{\partial^2 w}{\partial y^2} + 2N_{xy} \frac{\partial^2 w}{\partial x \, \partial y} = 0. \qquad (3.6)$$

In deriving Eq. (3.6), the weight of the plate has not been taken into consideration. We

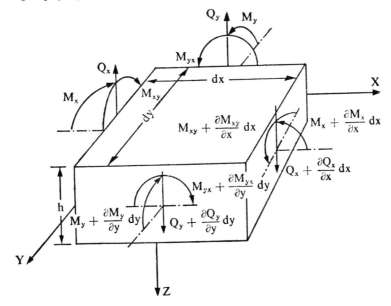

Fig. 3.4   Moment and Shear Forces on Plate Element.

take the moments of all the forces acting on the element ($dx \, dy$) about the X-axis and neglect small quantities of an order higher than those retained to obtain

$$\frac{\partial M_{xy}}{\partial x} - \frac{\partial M_y}{\partial y} + Q_y = 0. \qquad (3.7)$$

Similarly, by taking moments about the Y-axis, we get

$$\frac{\partial M_{yx}}{\partial y} + \frac{\partial M_x}{\partial x} - Q_x = 0. \tag{3.8}$$

Differentiating Eqs. (3.7) and (3.8) with respect to y and x, respectively, and noting that $M_{yx} = -M_{xy}$, we can express (3.6) as

$$\frac{\partial^2 M_x}{\partial x^2} - 2\frac{\partial^2 M_{xy}}{\partial x \, \partial y} + \frac{\partial^2 M_y}{\partial y^2} + N_x \frac{\partial^2 w}{\partial x^2} + N_y \frac{\partial^2 w}{\partial y^2} + 2N_{xy} \frac{\partial^2 w}{\partial x \, \partial y} = 0. \tag{3.9}$$

To obtain the governing equation in terms of the transverse displacement w, the moments should be expressed in terms of displacements. This can be achieved by relating the moments to stresses [as in Eq. (3.3)], and the stresses to strains and the strains to displacements (which we shall now do).

The stress-strain relation for a plate with elastic behaviour is

$$\sigma_x = \frac{E}{1 - \nu^2}(\epsilon_x + \nu\epsilon_y), \tag{3.10}$$

$$\sigma_y = \frac{E}{1 - \nu^2}(\epsilon_y + \nu\epsilon_x), \tag{3.11}$$

$$\tau_{xy} = \frac{E}{2(1 + \nu)}\gamma_{xy}, \tag{3.12}$$

where ν is Poisson's ratio.

In view of the two assumptions stated in Section 3.2, the displacement u, along the X-direction, of a point at a distance z above the middle surface of the plate is given by

$$u = -z\frac{\partial w}{\partial x}. \tag{3.13}$$

Similarly, in the Y-direction,

$$v = -z\frac{\partial w}{\partial y}. \tag{3.14}$$

Hence, in terms of the displacements, the strains ϵ_x, ϵ_y, and γ_{xy} are defined by

$$\epsilon_x = -z\frac{\partial^2 w}{\partial x^2}, \qquad \epsilon_y = -z\frac{\partial^2 w}{\partial y^2}, \qquad \gamma_{xy} = -2z\frac{\partial^2 w}{\partial x \, \partial y}. \tag{3.15}$$

Substituting Eqs. (3.15) in Eqs. (3.10)–(3.12), we obtain the stress-curvature relations

$$\sigma_x = -\frac{Ez}{1 - \nu^2}\left(\frac{\partial^2 w}{\partial x^2} + \nu\frac{\partial^2 w}{\partial y^2}\right), \tag{3.16a}$$

$$\sigma_y = -\frac{Ez}{1 - \nu^2}\left(\frac{\partial^2 w}{\partial y^2} + \nu\frac{\partial^2 w}{\partial x^2}\right), \tag{3.16b}$$

$$\tau_{xy} = -\frac{Ez}{1 + \nu}\frac{\partial^2 w}{\partial x \, \partial y}. \tag{3.16c}$$

Finally, by substituting Eqs. (3.16) in Eqs. (3.3b) and (3.3c) and integrating, we get the moment-curvature relations

$$M_x = -D\left(\frac{\partial^2 w}{\partial x^2} + \nu\frac{\partial^2 w}{\partial y^2}\right), \tag{3.17a}$$

$$M_y = -D(\frac{\partial^2 w}{\partial y^2} + \nu \frac{\partial^2 w}{\partial x^2}), \tag{3.17b}$$

$$M_{xy} = D(1 - \nu) \frac{\partial^2 w}{\partial x\, \partial y}, \tag{3.17c}$$

where D is the flexural rigidity per unit length of the plate and is given by

$$D = \frac{Eh^3}{12(1 - \nu^2)}. \tag{3.17d}$$

The flexural rigidity D is analogous to the bending stiffness EI of a beam. A comparison of the rigidity of the beam with that of the plate indicates that a strip of plate is stiffer than a beam of the same width and depth by a factor $1/(1 - \nu^2)$. This is so because a beam is free to deform laterally unlike a plate which is constrained due to the presence of adjacent strips.

By substituting Eqs. (3.17) in Eq. (3.9), we get

$$D\nabla^4 W = N_x \frac{\partial^2 w}{\partial x^2} + N_y \frac{\partial^2 w}{\partial y^2} + 2N_{xy} \frac{\partial^2 w}{\partial x\, \partial y}, \tag{3.18}$$

where

$$\nabla^4 W = \frac{\partial^4 w}{\partial x^4} + 2 \frac{\partial^4 w}{\partial x^2\, \partial y^2} + \frac{\partial^4 w}{\partial y^4}. \tag{3.19}$$

Equation (3.18) is the governing differential equation for the buckling of a uniform plate under the action of the inplane forces N_x, N_y, and N_{xy}.

For a nonuniform plate (plate thickness is a function of x and y and is symmetric with respect to the Z-axis), the process of deriving the governing equation is intricate. Here, the procedure changes after substituting (3.17) in (3.9) and at this step the derivatives of the stiffness quantity D have to be taken into account.

Example 3.1   Consider a simply-supported rectangular plate subjected to a uniform compressive loading N_x along the edges $x = 0$ and $x = a$ (see Fig. 3.5). In (3.18), by putting N_y and N_{xy}

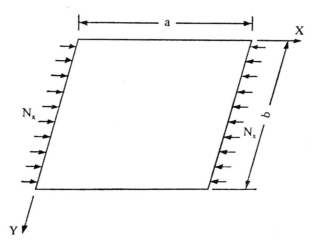

Fig. 3.5 Simply-Supported Plate Subjected to Uniform Compressive Loading.

equal to zero and $N_x = -N_x$, we reduce the governing equation for plate bending to

$$\frac{\partial^4 w}{\partial x^4} + 2\frac{\partial^4 w}{\partial y^2\,\partial x^2} + \frac{\partial^4 w}{\partial y^4} + N_x\frac{\partial^2 w}{\partial x^2} = 0. \tag{3.20}$$

The boundary conditions for the plate are

$$w = 0 = \frac{\partial^2 w}{\partial x^2} + v\frac{\partial^2 w}{\partial y^2} \qquad (x = 0,\,a),$$

$$w = 0 = \frac{\partial^2 w}{\partial y^2} + v\frac{\partial^2 w}{\partial y^2} \qquad (y = 0,\,b). \tag{3.21}$$

Since $w = 0$ along $x = 0$, a, we have $\partial^2 w/\partial y^2 = 0$. Similarly, $\partial^2 w/\partial x^2 = 0$ along $y = 0$, b. Hence, Eqs. (3.21) reduce to

$$\frac{\partial^2 w}{\partial x^2} = 0 \qquad (x = 0,\,a),$$

$$\frac{\partial^2 w}{\partial y^2} = 0 \qquad (y = 0,\,b). \tag{3.22}$$

The governing equation (3.20) and the boundary conditions (3.22) contain terms of even order in x and y. Hence, we can seek a variable-separable type of solution for (3.20). Let us assume that the solution is

$$w(x,\,y) = \sum_{n=1}^{\infty}\sum_{m=1}^{\infty} A_{mn}\sin\frac{m\pi x}{a}\sin\frac{n\pi y}{b} \qquad (m = 1,\,2,\,3,\,\ldots,\,n = 1,\,2,\,3,\,\ldots), \tag{3.23}$$

where m, n define the number of half waves that the plate buckles along the X- and Y-direction and A_{mn} represents the amplitude of the mode shapes. Equation (3.23) satisfies the boundary conditions (3:22) exactly. The substitution of the appropriate derivatives of $w(x,\,y)$ in (3.20) leads to

$$\sum_{m=1}^{\infty}\sum_{n=1}^{\infty} A_{mn}\left(\frac{m^4\pi^4}{a^4} + 2\frac{m^2 n^2\pi^4}{a^2 b^2} + \frac{n^4\pi^4}{b^4} - N_x\frac{m^2\pi^2}{a^2}\right)\sin\frac{m\pi x}{a}\sin\frac{n\pi y}{b} = 0. \tag{3.24}$$

In Eq. (3.24), the terms within the parentheses represent the sum of an infinite number of positive functions. In order that this sum be zero, the coefficient of every one of the terms should be equal to zero. That is,

$$A_{mn}\left[\pi^4\left(\frac{m^2}{a^2} + \frac{n^2}{b^2}\right)^2 - \frac{N_x}{D}\frac{m^2\pi^2}{a^2}\right] = 0. \tag{3.25}$$

Equation (3.25) is satisfied when (a) $A_{mn} = 0$ or (b) the expression within the square brackets is equal to zero. The first condition leads to a trivial case, i.e., the plate remains perfectly flat for all loads. The second condition results in a nontrivial solution leading to a critical load given by

$$(N_x)_{cr} = D\frac{\pi^2}{b^2}\left(\frac{mb}{a} + \frac{n^2 a}{mb}\right)^2. \tag{3.26}$$

Since we are interested in the lowest value of N_x satisfying Eq. (3.26), the values of m, n that minimize (3.26) must be determined. It is evident that N_x increases as n increases and that 1 is the minimum value of n, implying that there can be only one half wave along the Y-direction. Along the X-direction, the number of half waves corresponding to a minimum value of N_x is found by taking the derivative of (3.26) with respect to m and by putting it equal to zero. Thus,

$$\frac{d(N_x)}{dm} = \frac{2D\pi^2}{b^2}(\frac{mb}{a} + \frac{a}{mb})(\frac{b}{a} - \frac{a}{bm^2}) = 0.$$

This yields $m = a/b$. Substituting the value of m in (3.26), we get

$$(N_x)_{cr} = 4D\pi^2/b^2. \tag{3.27}$$

This relation is valid if a/b is an integer. For noninteger values, Eq. (3.27) becomes

$$(N_x)_{cr} = kD\pi^2/b^2, \tag{3.28}$$

where

$$k = (\frac{mb}{a} + \frac{a}{mb})^2. \tag{3.29}$$

Figure 3.6 shows the plot of K for the various values of a/b with integer values of m. It should be observed that the square plate has the lowest critical load. Further, for integer values of a/b,

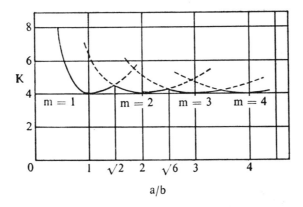

Fig. 3.6   Variation of Buckling Coefficient K with Aspect Ratio.

the plate divides into square plates when it buckles. For noninteger values of a/b, the buckling load of the plate is higher than that for the square plate. As long as $a/b < 1$, K varies considerably with the aspect ratio. However, for $a/b > 4$, the variation of K is negligible and, hence, a fairly accurate value of K is 4.0 for $a/b > 4.0$. The critical stress σ_x is obtained as

$$(\sigma_x)_{cr} = \frac{k\pi^2 E}{12(1 - \nu^2)}(\frac{h}{b})^2, \tag{3.30}$$

$$(\sigma_x)_{cr} = KE(h/b)^2. \tag{3.31}$$

The critical stress for the plate depends on its width, whereas for columns it varies with the length.

The method of solution becomes complicated if all the edges of the plate are not supported.

3.3.1   BUCKLING OF UNIFORM RECTANGULAR PLATE

Let us consider a simply-supported rectangular plate subjected to a uniform compressive loading N_x expressed per unit length along the edges $x = 0$ and $x = a$ (see Fig. 3.7).

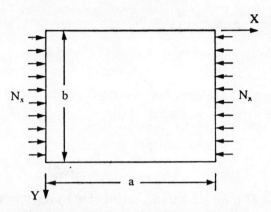

Fig. 3.7   Uniform Rectangular Plate under Compressive Loading.

The governing equation for thin plates is given by (3.20). Since the boundary along $x = 0$, a is simply-supported, a one-term solution of the type

$$w(x, y) = g(y) \sin \frac{m\pi x}{a} \tag{3.32}$$

satisfies the boundary conditions exactly. The unknown constant is contained in the function $g(y)$. The substitution of Eq. (3.32) in (3.20) leads to the ordinary differential equation

$$\frac{d^4 g}{dy^4} - \frac{2m^2 \pi^2}{a^2} \frac{d^2 g}{dy^2} + \left(\frac{m^4 \pi^4}{a^4} - \frac{N_x}{D} \frac{m^2 \pi^2}{a^2}\right) g = 0. \tag{3.33}$$

The characteristic equation of Eq. (3.33) has two real and two complex roots, namely,

$$\alpha = \pm \left[\frac{m^2 \pi^2}{a^2} + \left(\frac{N_x}{D} \frac{m^2 \pi^2}{a^2}\right)^{1/2}\right]^{1/2}, \qquad \beta = \pm \left[\frac{-m^2 \pi^2}{a^2} + \left(\frac{N_x}{D} \frac{m^2 \pi^2}{a^2}\right)^{1/2}\right]^{1/2}. \tag{3.34}$$

The solution to (3.33) can then be expressed in the form

$$g(y) = A e^{-\alpha y} + B e^{\alpha y} + C \cos \beta y + D \sin \beta y, \tag{3.35}$$

where A, B, C, and D are the constants and have to be determined from the boundary conditions along the edges $y = 0$ and $y = b$. For example, if the edge $y = 0$ is simply-supported and the edge $y = b$ is free, then

$$w = 0, \quad \frac{\partial^2 w}{\partial y^2} + \nu \frac{\partial^2 w}{\partial x^2} = 0 \qquad (y = 0), \tag{3.36}$$

$$\frac{\partial^2 w}{\partial y^2} + \nu \frac{\partial^2 w}{\partial x^2} = 0, \quad \frac{\partial^3 w}{\partial y^3} + (2 - \nu) \frac{\partial^3 w}{\partial x^2 \, \partial y} = 0 \qquad (y = b). \tag{3.37}$$

For these boundary conditions, the characteristic equation is

$$\beta \left(\alpha^2 - \nu \frac{m^2 \pi^2}{a^2}\right)^2 \tanh \alpha b = \alpha \left(\beta^2 + \nu \frac{m^2 \pi^2}{a^2}\right)^2 \tan \beta b. \tag{3.38}$$

The critical load N_x is contained in the expression for α and β and the critical stress σ_x is

$$(\sigma_x)_{cr} = k \frac{\pi^2 D}{b^2 h}. \tag{3.39}$$

Similarly, we can build up the solution for other boundary conditions.

3.4 ENERGY APPROACH

The differential equation approach becomes too intricate for nonuniform plates (as observed in respect of columns) and also for situations where boundary constraints are complex. In such instances, the energy method leads to a sound approximate solution for buckling loads and stresses.

In Chapter 2, we used the energy method to determine the critical load of a column by considering only the bending energy due to the bending stress σ_x. For a plate, the same method may be applied to derive the critical load, but here the strain energy due to all possible internal stresses is also taken into account.

For a thin plate, the possible stresses are σ_x, σ_y, and τ_{xy} and the total strain energy stored is given by

$$U = \tfrac{1}{2} \int_v (\sigma_x \epsilon_x + \sigma_y \epsilon_y + \tau_{xy}\gamma_{xy}) \, dx \, dy \, dz. \tag{3.40}$$

Making use of the stress-strain relations (3.10)–(3.12) and the relation between the stresses and transverse displacements (3.16a)–(3.16c), we can express Eq. (3.40), after integrating with respect to z, as

$$U = \frac{D}{2} \int_0^a \int_0^b [(\frac{\partial^2 w}{\partial x^2} + \frac{\partial^2 w}{\partial y^2})^2 - 2(1 - \nu)\{\frac{\partial^2 w}{\partial x^2} \frac{\partial^2 w}{\partial y^2} - (\frac{\partial^2 w}{\partial x \, \partial y})^2\}] \, dx \, dy. \tag{3.41}$$

Considering the loads as acting one at a time and then adding the components of the potential energy, we obtain the potential energy of the external loads as

$$V = -\tfrac{1}{2} \int_0^a \int_0^b [N_x(\frac{\partial w}{\partial x})^2 + N_y(\frac{\partial w}{\partial y})^2 + 2N_{xy} \frac{\partial w}{\partial x} \frac{\partial w}{\partial y}] \, dx \, dy. \tag{3.42}$$

The total potential energy π_p is given by

$$\pi_p = U + V. \tag{3.43}$$

At the point of buckling, π_p has a stationary value. Hence,

$$\partial(\pi_p) = \partial(U + V) = 0. \tag{3.44}$$

Example 3.2   Consider a plate which is fixed along two edges, simply-supported along the other two edges, and subjected to a uniform load N_x per unit length (see Fig. 3.8). The boundary

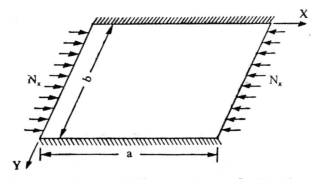

Fig. 3.8   Uniform Plate under Compressive Loading.

conditions are

$$w = \frac{\partial w}{\partial y} = 0 \qquad (y = 0, b),$$

$$w = \frac{\partial^2 w}{\partial x^2} + \nu\frac{\partial^2 w}{\partial y^2} = 0 \qquad (x = 0, a).$$

(3.45)

A one-term solution of the type

$$w(x, y) = A \sin\frac{m\pi x}{a}(1 - \cos\frac{2\pi y}{b}) \tag{3.46}$$

satisfies the boundary conditions (3.45) exactly. We have noted in Example 3.1 that the plate, regardless of its size, has one half wave along the Y-direction at the time of buckling. The substitution of Eq. (3.46) in Eqs. (3.41) and (3.42) and subsequent integration lead to

$$U = \frac{D}{2}A^2\frac{ab}{4}\frac{4\pi^2}{b^2} + \frac{m^2\pi^2}{a^2})^2 + \frac{2m^4\pi^4}{a^4}, \tag{3.47}$$

$$V = \tfrac{3}{8}N_x A^2 ab\frac{m^2\pi^2}{a^2}. \tag{3.48}$$

Therefore,

$$U + V = \frac{DA^2}{2}\frac{ab}{4}[(\frac{4\pi^2}{b^2} + \frac{m^2\pi^2}{a^2})^2 + \frac{2m^4\pi^4}{a^4}] - \tfrac{3}{8}N_x A^2 ab\frac{m^2\pi^2}{a^2}. \tag{3.49}$$

By putting

$$\frac{\partial}{\partial A}(U + V) = 0,$$

we get

$$(N_x)_{\text{cr}} = \tfrac{1}{3}\frac{Da^2}{m^2\pi^2}[(\frac{4\pi^2}{b^2} + \frac{m^2\pi^2}{a^2})^2 + \frac{2m^4\pi^4}{a^4}]. \tag{3.50}$$

From the result obtained in Example 3.2, we can see that, for a square plate with $m = 1$, the critical load is

$$(N_x)_{\text{cr}} = (7.69)\frac{\pi^2 D}{a^2}. \tag{3.51}$$

3.5 BUCKLING OF STIFFENED PLATES

Whenever large areas have to be roofed without intermediate columns, a form of construction known as *grid floor* is adopted. In this, a slab is stiffened by beams intersecting each other at right angles. In aircraft construction, the aircraft skin is normally stiffened along either one direction or both directions. Such types of construction are said to be structurally orthotropic. Another type of orthotropy is material orthotropy such as wood and composite materials. Stiffened plates may be treated in the same manner as orthotropic plates because they too have different rigidities along the two orthogonal directions. The method of treatment is simple and gives fairly accurate results.

For the buckling of an orthotropic plate, the moment relations (3.9) and the strain-displacement relations (3.15) remain the same as in isotropic plate buckling. However, the elastic properties differ in the two coordinate directions. The stress-strain relations are given by

$$\sigma_x = \frac{E_x}{(1 - v_x v_y)}(\epsilon_x + v_y \epsilon_y), \tag{3.52}$$

$$\sigma_y = \frac{E_y}{(1 - v_x v_y)}(\epsilon_y + v_x \epsilon_x), \tag{3.53}$$

$$\tau_{xy} = G_{xy}\gamma_{xy} = \frac{(E_x E_y)^{1/2}}{2[1 - (v_x v_y)^{1/2}]}\gamma_{xy}, \tag{3.54}$$

where E_x, E_y are the elastic constants along the X- and Y-direction and v_x, v_y are Poisson's ratios along the X- and Y-direction. Applying Betti's reciprocal theorem, we get the relationship between the two elastic constants as

$$v_y E_x = v_x E_y. \tag{3.55}$$

With the stress-strain relations defined by Eqs. (3.52), (3.53), and (3.54), the moment relations (3.9) may be rewritten as

$$.M_x = -D_x\left(\frac{\partial^2 w}{\partial x^2} + v_y\frac{\partial^2 w}{\partial y^2}\right), \tag{3.56}$$

$$M_y = -D_y\left(\frac{\partial^2 w}{\partial y^2} + v_x\frac{\partial^2 w}{\partial x^2}\right), \tag{3.57}$$

$$M_{xy} = -2D_{xy}\frac{\partial^2 w}{\partial x\ \partial y}, \tag{3.58}$$

where

$$D_x = \frac{E_x h^3}{12(1 - v_x v_y)}, \tag{3.59a}$$

$$D_y = \frac{E_y h^3}{12(1 - v_x v_y)}, \tag{3.59b}$$

$$D_{xy} = G_{xy}\frac{h^3}{12}. \tag{3.59c}$$

The governing equation for the buckling of the orthotropic plate is then expressed as

$$D_x\frac{\partial^4 w}{\partial x^4} + 2H\frac{\partial^4 w}{\partial x^2\ \partial y^2} + D_y\frac{\partial^4 w}{\partial y^4} = N_x\frac{\partial^2 w}{\partial x^2} + N_y\frac{\partial^2 w}{\partial y^2} + 2N_{xy}\frac{\partial^2 w}{\partial x\ \partial y}. \tag{3.60}$$

The critical load is obtained in the same manner as described in Section 3.3.

3.6 NUMERICAL TECHNIQUES

If a plate has well-defined boundaries, the techniques described in Sections 3.3–3.5 can be usefully applied to obtain a closed-form solution. In practice, however, there are always discontinuities at discrete points in a structure and it is futile to seek a closed-form solution. Since the plate problem is two-dimensional, the application of the finite difference and finite element methods is more involved; nevertheless, in such situations, these techniques prove useful from the computation point of view.

3.6.1 FINITE DIFFERENCE TECHNIQUE

In Section 2.7.1, we used the finite difference technique to obtain the buckling load of a column. To extend this method to plates, it is necessary to evolve expressions for difference ratios corresponding to partial derivatives. Let us consider a plate represented by a network of discrete points, evenly placed at a distance h from each other, along the X- and Y-direction, as shown in Fig. 3.9. The distance along x and y need not however be the same. The second

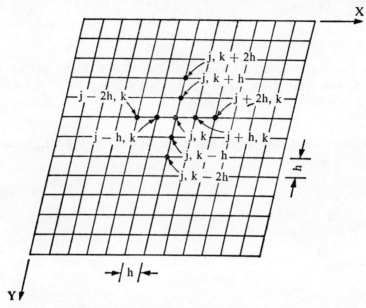

Fig. 3.9   Finite Difference Mesh for Plate.

and fourth differences with respect to x are

$$(\frac{\partial^2 w}{\partial x^2})_{i,j} = (w_{i+h,j} - 2w_{i,j} + w_{i-h,j})/h^2, \tag{3.61}$$

$$(\frac{\partial^4 w}{\partial x^4})_{i,j} = (w_{i+2h,j} - 4w_{i+h,j} + 6w_{i,j} - 4w_{i-h,j} + w_{i-2h,j})/h^4. \tag{3.62}$$

Similarly, the second and fourth differences with respect to y are

$$(\frac{\partial^2 w}{\partial y^2})_{i,j} = (w_{i,j+h} - 2w_{i,j} + w_{i,j-h})/h^2, \tag{3.63}$$

$$(\frac{\partial^4 w}{\partial y^4})_{i,j} = (w_{i,j+2h} - 4w_{i,j+h} + 6w_{i,j} - 4w_{i,j-h} + w_{i,j-2h})/h^4, \tag{3.64}$$

$$(\frac{\partial^4 w}{\partial x^2 \partial y^2})_{i,j} = \frac{\partial^2}{\partial x^2}(\frac{\partial^2 w}{\partial y^2})_{i,j} \equiv \frac{1}{h^2}[(\frac{\partial^2 w}{\partial y^2})_{i+h,j} - 2(\frac{\partial^2 w}{\partial y^2})_{i,j} + (\frac{\partial^2 w}{\partial y^2})_{i-h,j}], \tag{3.65}$$

$$(\frac{\partial^4 w}{\partial x^2 \partial y^2})_{i,j} \approx \frac{1}{h^4}[(w_{i+h,j+h} - 2w_{i+h,j} + w_{i+h,j-h}) - 2(w_{i,j+h} - 2w_{i,j} + w_{i,j-h})$$
$$+ (w_{i-h,j+h} - 2w_{i-h,j} + w_{i-h,j-h})]. \tag{3.66}$$

Example 3.3   Consider a square plate of side a, simply-supported along two opposite edges, clamped along the other two edges, and subjected to a uniform compressive loading N_x (see Fig. 3.10). Taking the coordinate axes as shown, we obtain the differential equation

$$\nabla^4 w + \frac{N_x}{D} \frac{\partial^2 w}{\partial x^2} = 0. \tag{3.67}$$

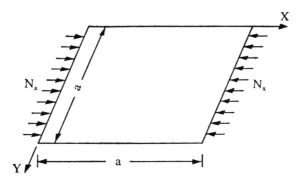

Fig. 3.10   Square Plate under Uniform Compressive Loading.

Further, the boundary conditions are

$$w = 0, \quad \frac{\partial^2 w}{\partial x^2} = 0 \qquad (x = 0, a),$$

$$w = 0, \quad \frac{\partial w}{\partial y} = 0 \qquad (y = 0, a). \tag{3.68}$$

Introducing the nondimensional parameters ξ, η, defined as

$$\xi = x/a, \qquad \eta = y/a, \tag{3.69}$$

we can rewrite (3.67) as

$$\nabla^4 w + \frac{N_x a^2}{D} \frac{\partial^2 w}{\partial \xi^2} = 0. \tag{3.70}$$

If the plate is covered with a square lattice of mesh size $1/n$, then by multiplying Eq. (3.70) by $h^4 = 1/n^4$, and writing the differences for the second and fourth derivatives, we have

$$
\begin{aligned}
w_{i+2, j} &- 8w_{i+1, j} + 20w_{i, j} - 8w_{i-1, j} + w_{i-2, j} + 2w_{i+1, j+1} + 2w_{i+1, j-1} - 8w_{i, j+1} \\
&- 8w_{i, j-1} + w_{i, j+2} + w_{i, j-2} + 2w_{i-1, j+1} + 2w_{i-1, j-1} \\
&+ \lambda_n(w_{i+1, j} - 2w_{i, j} + w_{i-1, j}) = 0, \quad (3.71)
\end{aligned}
$$

where $\lambda_n = N_x a^2/(Dn^2)$. The application of Eq. (3.71) at the internal pivotal points $(i, j = 1, 2, \ldots, n - 1)$ gives rise to a set of homogeneous, algebraic equations in the unknown pivotal displacements $w_{i, j}$ whose determinant is identically equal to zero for a nontrivial solution. The lowest value of N_x, which renders the determinant equal to zero, gives the first critical load. Writing the boundary conditions (3.68) in terms of differences, we have

$$
\begin{aligned}
w_{i, j} &= 0, \quad & w_{1, j} &= -w_{-1, j}, \quad & w_{n+1, j} &= w_{n-1, j}, \\
w_{i, j} &= 0, \quad & w_{i, n+1} &= w_{i, n-1}, \quad & w_{i, 1} &= -w_{i, -1}.
\end{aligned}
\tag{3.72}
$$

For $n = 3$, $\lambda_3 = 6$. Therefore,

$$(N_x)_{\mathrm{cr}} = 54 \frac{D}{a^2} = (5.47)\pi^2 \frac{D}{a^2}. \tag{3.73}$$

For $n = 4$, $\lambda_4 = 3.82$. Hence,

$$(N_x)_{\mathrm{cr}} = (61.1)\frac{D}{a^2} = (6.19)\pi^2 \frac{D}{a^2}. \tag{3.74}$$

The number of meshes can be increased, depending on the desired accuracy of $(N_x)_{\mathrm{cr}}$.

3.6.2   FINITE ELEMENT TECHNIQUE

The finite element technique developed in Section 2.7.2 for determining the critical load of a column is also applicable to the solution of plate buckling problems. As in the case of the column, the plate too is subdivided into a large number of discrete elements connected at the nodal points. But whereas only beam elements are considered in the case of the column (because the column is one-dimensional), for a plate (which is two-dimensional) a variety of elements—such as triangular, rectangular, trapezoidal, and sectorial—are possible. The choice of element depends.on the shape as also the boundary of the plate. For example, for a circular plate, a sectorial or triangular element is preferred since its boundaries can be easily matched. For a rectangular plate, a rectangular element is better suited since it can cover the plate exactly. (For an exhaustive treatment of this topic, see the bibliography at the end of this chapter.)

In general, we can have two coordinate systems: the *global coordinate system*, which is fixed and with respect to which the displacements, slopes, and moments are measured; and the *local coordinate system*, which changes from one element to another. The relationship between these two systems is expressed in terms of their direction cosines. The local coordinate system is usually preferred in view of the fewer number of variables. For example, let us consider an axial bar element lying in a plane; if the local coordinate axis lies along the axis of the bar, only two axial displacements of the bar are possible. However, for any other axis, there are two displacements at each node, thus increasing the order of the matrix.

As in the case of columns, the geometric stiffness matrix for plates too may be consistent or inconsistent. The displacement curve is now a function of two variables (x, y). A displacement function which is continuous within the element and compatible between adjacent elements is termed *compatible* or *conforming*. A variety of displacement functions is available, some conformable and others nonconformable. It is rather difficult to obtain the conformable type of elements. The convergence of the solution would be monotonic if a conformable type of displacement function is used.

Example 3.4   Consider a rectangular plate of uniform thickness, subjected to an inplane loading, as shown in Fig. 3.11. The plate is divided into a large number of rectangular el-

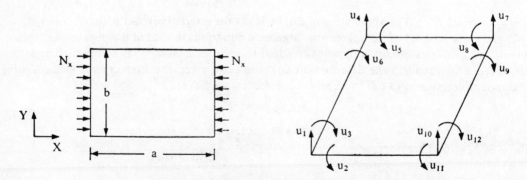

Fig. 3.11   Loaded Rectangular Plate and Nodal Degrees of Freedom.

ements. At each nodal point, we define three displacements. These are a vertical deflection w and rotations $\partial w/\partial x$ and $\partial w/\partial y$ about the X- and Y-axis. Thus, there are 12 degrees of freedom for each element. To evaluate the elastic stiffness matrix $[k]$, we assume that the deflection w

at any point on the element can be represented by a 12-term polynomial. Thus,

$$w = \alpha_1 + \alpha_2 x + \alpha_3 y + \alpha_4 x^2 + \alpha_5 xy + \alpha_6 y^2 + \alpha_7 x^3 + \alpha_8 x^2 + \alpha_9 xy^2 + \alpha_{10} y^3$$
$$+ \alpha_{11} x^3 y + \alpha_{12} xy^3, \tag{3.75}$$

where $\alpha_1, \alpha_2, \ldots, \alpha_{12}$ are the generalized coordinates. In terms of the nondimensional parameters ξ and η, defined as $\xi = x/a$, $\eta = y/b$, the matrix $[N]$ can be obtained [see Eq. (2.88)] as

$$N^T = \begin{bmatrix} 1 - \xi\eta - (3 - 2\xi)\xi^2(1 - \eta) - (1 - \xi)(3 - 2\eta)\eta^2 \\ (1 - \xi)\eta(1 - \eta)^2 b \\ -\xi(1 - \xi)^2(1 - \eta)a \\ (1 - \xi)(3 - 2\eta)\eta^2 + \xi(1 - \xi)(1 - 2\xi)\eta \\ -(1 - \xi)(1 - \eta)\eta^2 b \\ -\xi(1 - \xi)^2\eta a \\ (3 - 2\xi)\xi^2\eta - \xi\eta(1 - \eta)(1 - 2\eta) \\ -\xi(1 - \eta)\eta^2\beta \\ (1 - \xi)\xi^2\eta a \\ (3 - 2\xi)\xi^2(1 - \eta) + \xi\eta(1 - \eta)(1 - 2\eta) \\ \eta\xi(1 - \eta^2)b \\ (1 - \xi)\xi^2(1 - \eta)a \end{bmatrix} \tag{3.76}$$

Evidently, the displacement at any point along the edge depends entirely on the nodal displacements of the edge. In view of this, the boundary deflections on adjacent plate elements are compatible. However, rotations of the element edges on a common boundary are not compatible, and hence discontinuities in the slopes exist across the boundaries.

The function chosen for w in Eq. (3.75) satisfies equilibrium and compatibility within the element. By following the procedure given in Section 2.7.2, we can obtain the elastic stiffness matrix and the geometric stiffness matrix for each element of the rectangular plate. The total stiffness and geometric matrices can be derived by adding the element stiffness matrices, and the eigenvalues can be obtained by solving the characteristic equation.

Today, there is no dearth of material dealing with the application of the finite element technique to the problems in structural mechanics.

BIBLIOGRAPHY

Bathe, K. J. and Wilson, E., Numerical Methods in Finite Element Analysis, Prentice-Hall, Englewood Cliffs, New Jersey, 1976.

Bleich, F., Buckling Strength of Metal Structures, McGraw-Hill, New York, 1952.

Chajes, A., Principles of Structural Stability Theory, Prentice-Hall, Englewood Cliffs, New Jersey, 1974.

Desai, C. S. and Abel, J. F., Introduction to the Finite Element Method, Van Nostrand Reinhold, New York, 1972.

Livesley, R. K., Matrix Methods of Structural Analysis, Pergamon Press, Oxford, 1964.

Martin, H. C., Introduction to Matrix Methods of Structural Analysis, McGraw-Hill, New York, 1966.

Martin, H. C. and Carey, G. F., Introduction to Finite Element Analysis, McGraw-Hill, New York, 1973.

Meek, J. L., Matrix Structural Analysis, McGraw-Hill, New York, 1971.

Przemieniecki, J. S., Theory of Matrix Structural Analysis, McGraw-Hill, New York, 1968.

Ralston, A. and Wilf, H. S., Mathematical Methods for Digital Computers, Wiley, New York, 1965.

Rubinstein, M. F., Matrix Computer Analysis of Structures, Prentice-Hall, Englewood Cliffs, New Jersey, 1966.

Salvadori, M. G. and Baron, M. L., Numerical Methods in Engineering, Prentice-Hall, Englewood Cliffs, New Jersey, 1961.

Timoshenko, S. P. and Gere, J. M., Theory of Elastic Stability, McGraw-Hill, New York, 1961.

Zienkiewicz, O. C., The Finite Element Method in Engineering Science, 2nd edn., McGraw-Hill, London, 1971.

Zienkiewicz, O. C. and Holister, G. S., Recent Developments in Numerical and Experimental Methods, Wiley, New York, 1965.

Zienkiewicz, O. C. and Cheung, Y. K., The Finite Element in Structural and Continuum Mechanics, McGraw-Hill, New York, 1967.

4

Vibration of Beams

4.1 INTRODUCTION

Structures are often subjected to movable and time-varying loads termed *dynamic loads*. These loads are classified as follows.

Periodic load   This load repeats itself at regular intervals of time t, e.g., a pulsating load associated with an unbalanced machine running at a constant speed, a suspension bridge, an aircraft wing, and a chimney.

Random load   This load varies erratically with time, e.g., an impulsive load produced by blast.

Moving load   Here, the magnitude of the load may be constant but its position varies with time, e.g., a vehicle moving on a bridge, and the contact force between a pantograph and a current-carrying conductor.

The analysis of a structure subjected to dynamic loads is different from static loads for two reasons. Firstly, under dynamic loading, quantities such as stress and deflection vary with time. Therefore, instead of a single solution, it is essential to find a solution at each interval of time during the period considered or to determine at least the critical solution during the whole period. Secondly, the inertia loads are only a part of the load system that causes stresses and deflections. The total stresses and deflections in a structure are due to a combination of applied and inertia forces. Since the inertia forces depend on and contribute to the deflection of a structure, a direct solution of the problem is not possible.

In general, a structure such as a beam has infinite degrees of freedom. However, for all practical problems, only finite degrees of freedom are considered since higher harmonics seldom get excited. If the load is random, some of the higher harmonics may be excited, but in most structures, the response may be approximated by four or five modes.

A structure is analyzed by idealizing it. Such idealization depends on the nature of loading and the type of structure and its cross-sectional symmetry. One of the simplest idealizations is the single-degree-of-freedom system. For instance, to find the displacement of a landing gear for the transverse motion of an aircraft, the mass of the wing is lumped at the fuselage and the landing gear is represented by a spring-dashpot system. Again, the natural frequency of an overhead water tank can be obtained by idealizing it as a single-degree-of-freedom system. This would result in a very crude approximation if the structure is made up of a large number of elastic members. Since a large number of books deal with the single-degree-of-freedom system, we shall confine our study to the multidegree-of-freedom system or continuous system.

In our analysis, it is assumed that vibration is restricted to one plane; this is true when the structure has two axes of symmetry. Open sections whose centre of gravity does not coincide with the shear centre have coupled modes.

4.2 CONTINUOUS SYSTEMS

As already stated, a continuous system may be considered an assemblage of spring-mass systems in which the spring represents the stiffness devoid of mass, all the mass being lumped at discrete points. In other words, an infinite-degree-of-freedom system is converted into a finite-degree-of-freedom system. For a continuous system, we assume that the mass and elasticity are distributed. We also assume that the body is homogeneous and isotropic, and that the stress-strain relation is linear within the elastic limit. Since an infinite number of coordinates is necessary to specify the position of a point, such a system possesses an infinite degree of freedom.

In general, the free vibration mode is the sum of the principal modes. In such modes, every particle of the body undergoes simple harmonic motion and simultaneously passes through its equilibrium position.

4.2.1 TRANSVERSE VIBRATION OF BEAMS

To obtain the equation of motion for the lateral or transverse vibration of a beam, we assume that the beam is of closed cross-section and that the shear centre coincides with the centre of gravity. For this reason, we can treat torsional and flexural vibrations separately. For closed sections, torsional stiffness GJ is larger than bending or flexural stiffness EI, and hence we can expect the flexural mode to be predominant. Assuming small deflection, we may express the moment-curvature relation (see Fig. 4.1) as

$$EI\frac{\partial^2 v}{\partial z^2} = -M, \tag{4.1}$$

where $v(z, t)$ is the transverse displacement which is a function of the spatial coordinate z and the time t.

Figure 4.1 shows a differential element on which the shear force V, the bending moment

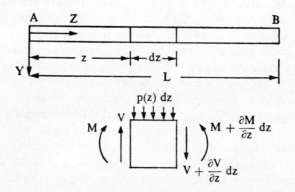

Fig. 4.1   Forces Acting on Beam Element.

M, and the inertia loading per unit length $p(z)$ are acting. The equilibrium of forces along the Y-direction yields

$$\frac{\partial V}{\partial z}\,dz + p(z)\,dz = 0. \tag{4.2}$$

The moment equilibrium gives

$$V \, dz + (\frac{\partial V}{\partial z} \, dz) \, dz - (\frac{\partial M}{\partial z} \, dz) - \tfrac{1}{2}p(z)(dz)^2 = 0. \tag{4.3}$$

From Eqs. (4.1), (4.2), and (4.3), we get

$$\frac{\partial^2}{\partial z^2}(EI \frac{\partial^2 v}{\partial z^2}) = -\frac{\partial V}{\partial z} = p(z). \tag{4.4}$$

For free vibration problems, $p(z)$ can be replaced by the inertia force through the use of d'Alembert's principle which is given by

$$p(z) = -\rho_b A_b \frac{\partial^2 v}{\partial t^2}, \tag{4.5}$$

where ρ_b is the density of the beam material and A_b the area of the cross-section of the beam.

For response problems, $p(z)$ contains applied load terms as well as inertia forces. Combining the two, we get the governing equation for free vibration as

$$\frac{\partial^2}{\partial z^2}(EI \frac{\partial^2 v}{\partial z^2}) + \rho_b A_b \frac{\partial^2 v}{\partial t^2} = 0. \tag{4.6}$$

If E and I are constants with respect to z, Eq. (4.6) reduces to

$$EI \frac{\partial^4 v}{\partial z^4} + \rho_b A_b \frac{\partial^2 v}{\partial t^2} = 0 \tag{4.7}$$

which is a partial, linear, and constant coefficient differential equation. Hence, we can seek a variable-separable type of solution. Let

$$v(z, t) = v(z)e^{i\omega t}, \tag{4.8}$$

where $v(z)$ is a function of the space coordinate only. Substituting Eq. (4.8) in (4.7), we obtain

$$d^4 v/dz^4 - \lambda_n^4 v = 0, \tag{4.9}$$

where $\lambda_n^4 = \rho_b A_b \omega_n^2 / EI$ and is called the frequency parameter.

The general solution of Eq. (4.9) is

$$v(z) = C_1 \cosh \lambda z + C_2 \sinh \lambda z + C_3 \cos \lambda z + C_4 \sin \lambda z, \tag{4.10}$$

where C_1, C_2, C_3, and C_4 are the constants to be obtained from the boundary conditions of the problem.

Example 4.1   For a simply-supported or hinged beam, the boundary conditions are

$$v(0) = 0, \qquad v(L) = 0,$$
$$EI \frac{\partial^2 v}{\partial z^2}(0) = 0, \qquad EI \frac{\partial^2 v}{\partial z^2}(L) = 0. \tag{4.11}$$

From (4.8), we can define the boundary conditions in terms of $v(z)$ alone. Using Eqs. (4.11) in Eq. (4.10), we reduce the characteristic equation for a nontrivial case to

$$\sin \lambda_n L = 0 \qquad (\lambda L = \pi, 2\pi, \ldots, n\pi). \tag{4.12}$$

Therefore,

$$\omega_n = \frac{\pi^2 n^2}{L^2}(\frac{\rho_b A_b}{EI})^{1/2},$$

$$f_n = \frac{\pi n^2}{2L^2}(\frac{\rho_b A_b}{EI})^{1/2}.$$

(4.13)

The period of vibration is

$$\tau_n = \frac{1}{f_n} = \frac{2L^2}{\pi n^2}(\frac{EI}{\rho_b A_b})^{1/2}.$$

(4.14)

Table 4.1 gives the values of λ for the first four modes for different boundary conditions. It is clear from this table that the hinged-free boundary has the lowest frequency. In practice, however, the boundary conditions lie between the ideal values, because the adjoining numbers are connected elastically to the supporting members.

Table 4.1   Values of λL for Various Boundary Conditions

Boundary conditions	$\lambda_1 L$	$\lambda_2 L$	$\lambda_3 L$	$\lambda_4 L$
Simply-supported	3.142	6.284	9.426	12.568
One end clamped, the other end free	1.875	4.694	7.855	10.996
Both ends free		4.732	7.853	10.996
Both ends clamped		4.732	7.853	10.996
One end clamped, the other end hinged	3.927	7.069	10.210	13.352
One end hinged, the other end free		3.927	7.069	10.210

If a member is short, i.e., if the cross-sectional dimensions are of the same order as the length, the effect of shear deformation cannot be neglected. Also, as the beam bends, the cross-sections rotate. For higher modes, such rotations, which give rise to rotatory inertia, are quite large. It can then be easily seen that the effective length of a beam is reduced, thus making the secondary effects important.

Figure 4.2 shows an element of a beam along with the forces acting on it. Due to shear,

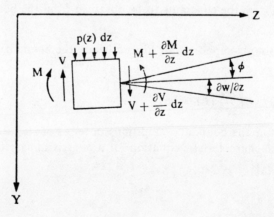

Fig. 4.2   Deformed Shape of Element with Forces.

the shape of the cross-section distorts though the face does not rotate. If ϕ is the shear angle, then the result is

$$\psi - \frac{\partial v}{\partial z} = -\frac{V}{k_s A_b G}, \tag{4.15}$$

$$\frac{\partial \psi}{\partial z} = -\frac{M}{EI}, \tag{4.16}$$

where ψ is the bending slope, k_s the shape factor, and G the shear modulus.

From the rotation of the element, we have

$$J_r \frac{\partial^2 \psi}{\partial t^2} = \frac{\partial M}{\partial z} - V, \tag{4.17}$$

$$m_b \frac{\partial^2 v}{\partial t^2} = -\frac{\partial V}{\partial z} - p(z, t), \tag{4.18}$$

where J_r and m_b are the rotatory inertia and mass per unit length of the beam, respectively. Substituting Eqs. (4.15) and (4.16) in Eqs. (4.17) and (4.18), we get

$$\frac{\partial}{\partial z}\left(EI \frac{\partial \psi}{\partial z}\right) + k_s A_b G\left(\frac{\partial v}{\partial z} - \psi\right) - J_r \frac{\partial^2 \psi}{\partial t^2} = 0, \tag{4.19}$$

$$m_b \frac{\partial^2 v}{\partial t^2} - \frac{\partial}{\partial z}\left(\psi - \frac{\partial v}{\partial z}\right) + p(z, t) = 0. \tag{4.20}$$

Eliminating ψ and assuming that the cross-section of the beam is uniform, we obtain

$$EI \frac{\partial^4 v}{\partial z^4} + m_b \frac{\partial^2 v}{\partial t^2} - \left(J_r + \frac{EI \cdot m_b}{k_s A_b G}\right)\frac{\partial^4 v}{\partial z^2 \partial t^2} + \frac{J_r \cdot m_b}{k_s A_b G}\frac{\partial^4 v}{\partial t^4}$$
$$+ \frac{J_r}{k_s A_b G}\frac{\partial^2 p}{\partial t^2} - \frac{EI}{k_s A_b G}\frac{\partial^2 p}{\partial z^2} = -p(z, t). \tag{4.21}$$

It should be observed that (4.7) is a special case of Eq. (4.21) which is referred to as the *Timoshenko beam*.

For the ideal configuration, the boundary conditions possible are

$$v = 0, \quad \frac{\partial \psi}{\partial z} = 0 \qquad \text{(simply-supported ends)},$$

$$v = 0, \quad \psi = 0 \qquad \text{(clamped ends)}, \tag{4.22}$$

$$\frac{\partial \psi}{\partial z} = 0, \quad \frac{\partial v}{\partial z} - \psi = 0 \qquad \text{(free ends)}$$

Assuming that the beam is undergoing harmonic oscillations, we can find the solution for $v(z, t)$ in terms of the nondimensional parameter $\xi \ (=z/L)$ as

$$v(\xi, t) = v(\xi)e^{i\omega t}. \tag{4.23}$$

For free vibrations, Eqs. (4.19), (4.20), and (4.21) may be expressed in terms of the non-dimensional parameter ξ as

$$K_s \frac{d^2 \psi}{d\xi^2} - (1 - \Omega^2 K_s^2 \mu^2)\psi + \frac{1}{L}\frac{dv}{d\psi} = 0, \tag{4.24}$$

$$\frac{d^2 v}{d\xi^2} + K_s^2 \Omega^2 v - L \frac{d\psi}{d\xi} = 0, \tag{4.25}$$

$$\frac{d^4v}{d\xi^4} + \Omega^2(\mu^2 + K_s^2)\frac{d^2v}{d\xi^2} - \Omega^2(1 - \Omega^2 K_s^2 \mu^2)v = 0, \tag{4.26}$$

where

$$\Omega^2 = \frac{\rho_b}{EI}A_b L^4 \omega^2, \qquad \mu^2 = \frac{I}{A_b L^2}, \qquad K_s^2 = \frac{EI}{k_s A_b G L^2}. \tag{4.27}$$

Here, the dimensionless parameter Ω is directly related to the natural frequency ω and the parameters μ and K_s are measures of the effects of rotatory inertia and shear deformation, respectively.

The solution to Eq. (4.26) may be written as

$$v(\xi) = C_1 \cosh \alpha\xi + C_2 \sinh \alpha\xi + C_3 \cos \beta\xi + C_4 \sin \beta\xi, \tag{4.28}$$

where

$$\alpha = \frac{1}{\sqrt{2}}\{[\{\Omega^2(\mu^2 - K_s^2)\}^2 + 4\Omega^2]^{1/2} - \Omega^2(\mu^2 + K_s^2)\}^{1/2},$$

$$\beta = \frac{1}{\sqrt{2}}\{[\{\Omega^2(\mu^2 - K_s^2)\}^2 + 4\Omega^2]^{1/2} + \Omega^2(\mu^2 + K_s^2)\}^{1/2}. \tag{4.29}$$

Equation (4.28) is the solution of (4.26), provided

$$[\Omega^2(\mu^2 - K_s^2)]^2 + 4\Omega^2 > 0, \tag{4.30a}$$

$$[\{\Omega^2(\mu^2 - K_s^2)\}^2 + 4\Omega^2]^{1/2} \pm \Omega^2(\mu^2 + K_s^2) = 0. \tag{4.30b}$$

Inequality (4.30a) is required so that α and β do not become complex. Since the real frequencies alone are of consequence, we consider only those values of Ω for which the inequality is satisfied. Any violation of Eq. (4.30b) means that either α or β is purely imaginary. Hence, these roots may be written as

$$\alpha = \frac{i}{\sqrt{2}}\{-[\{\Omega^2(\mu^2 - K_s^2)\}^2 + 4\Omega^2]^{1/2} - \Omega^2(\mu^2 + K_s)\}^{1/2} = i\alpha',$$

$$\beta = \frac{i}{\sqrt{2}}\{-[\{\Omega^2(\mu^2 - K_s^2)\}^2 + 4\Omega^2]^{1/2} + \Omega^2(\mu^2 + K_s^2)\}^{1/2} = i\beta'. \tag{4.31}$$

For α imaginary, we have

$$v(\xi) = C_1 \cos \alpha'\xi + C_2 \sin \alpha'\xi + C_3 \cos \beta\xi + C_4 \sin \beta\xi. \tag{4.32}$$

For β imaginary, we obtain

$$v(\xi) = C_1 \cos \alpha\xi + C_2 \sin \alpha\xi + C_3 \cosh \beta'\xi + C_4 \sinh \beta'\xi. \tag{4.33}$$

Using the boundary conditions (4.22), we get, for a simply-supported beam, the characteristic equation

$$2 - 2 \cosh \alpha \cos \beta - (\frac{\eta^2 - \zeta^2}{\eta\zeta}) \sinh \alpha \sin \beta = 0 \tag{4.34}$$

and the associated mode shape

$$v(\xi) = \cosh \alpha\xi - \cos \beta\xi - (\frac{\cosh \alpha - \cos \beta}{\sinh \alpha + \eta \sin \beta})(\zeta \sinh \alpha\xi + \eta \sin \beta\xi), \tag{4.35}$$

where

$$\eta = \frac{K_s^2 \Omega^2 + \alpha^2}{\alpha}, \qquad \zeta = \frac{K_s^2 \Omega^2 - \beta^2}{\beta}. \tag{4.36}$$

Given the cross-sectional dimensions and the length of a beam, it is now possible to ascertain the natural frequency and the mode shapes.

4.2.2   VIBRATION OF VISCOELASTIC BEAMS

So far we have considered the material of a beam to be elastic. This is true of metals at room temperature. At higher temperatures, they behave in the manner of viscoelastic material. Polymers which are finding a place in structural applications are viscoelastic even at room temperature. For such materials, the material property is a function of time. We confine our study to linear viscoelastic behaviour, since the correspondence principle, by means of which a viscoelastic problem is converted into an equivalent elastic problem, can be applied to it. The dependence on time is eliminated by the use of the Laplace transform.

For forced vibration of a beam of perfectly elastic material, (4.7) may be rewritten as

$$\beta^2 \frac{\partial^4 v}{\partial z^4} + \frac{\partial^2 v}{\partial t^2} = \frac{q(z, t)}{\rho_b A_b}, \tag{4.37}$$

where

$$\beta = \left(\frac{EI}{\rho_b A_b}\right)^{1/2}.$$

Applying the Laplace transform to Eq. (4.37), we obtain

$$\beta^2(p) \frac{d^4 \bar{v}}{dz^4} + p^2 \bar{v} = \frac{\bar{q}(z, t)}{\rho_b A_b} + pv(z, 0) + \frac{dv}{dt}(z, 0), \tag{4.38}$$

where

$$\bar{v}(z, p) = \int_0^\infty e^{-pt} v(z, t) \, dt. \tag{4.39}$$

The quantity $\bar{E}(p)$ contained in $\beta(p)$ can be determined as follows. In the transverse vibration of a beam, when the effect of shear is neglected, the state of stress at a point is given by

$$\sigma_{11} = E\epsilon_{11}, \qquad \sigma_{12} = \sigma_{13} = \sigma_{22} = \sigma_{23} = \sigma_{33} = 0. \tag{4.40}$$

In terms of Lamé's constants λ and μ, the generalized Hooke's law for elastic material is expressed as

$$\sigma_{ij} = \lambda_{ij}\epsilon_{kk} + 2\mu\epsilon_{ij} \qquad (i, j = 1, 2, 3). \tag{4.41}$$

Taking the Laplace transform on both sides, we may write Eqs. (4.41) for a viscoelastic material as

$$\bar{\sigma}_{11} = \bar{\lambda}(p)\bar{e} + 2\bar{\mu}(p)\epsilon_{11},$$
$$\bar{\sigma}_{22} = 0 = \bar{\lambda}(p)\bar{e} + 2\bar{\mu}(p)\epsilon_{22}, \tag{4.42}$$
$$\bar{\sigma}_{33} = 0 = \bar{\lambda}(p)\bar{e} + 2\bar{\mu}(p)\epsilon_{33},$$

where $\bar{e} = \epsilon_{11} + \epsilon_{22} + \epsilon_{33}$. The sum of the normal stresses yields

$$\bar{\sigma} = \bar{\sigma}_{11} = [3\bar{\lambda}(p) + 2\bar{\mu}(p)]\bar{e}.$$

Hence,

$$\bar{e} = \frac{\bar{\sigma}_{11}}{3\bar{\lambda}(p) + 2\bar{\mu}(p)}.$$

Substituting the value of $\bar{e}$ in relations (4.42), we obtain

$$\bar{\sigma}_{11} = \frac{\bar{\mu}(p)[3\bar{\lambda}(p) + 2\bar{\mu}(p)]}{\bar{\lambda}(p) + \bar{\mu}(p)}\bar{\epsilon}_{11} = \bar{E}(p)\bar{\epsilon}_{11}. \tag{4.43}$$

Let the initial conditions for forced vibration be

$$v(z, 0) = f(z), \qquad \frac{\partial v}{\partial t}(z, 0) = g(z). \tag{4.44}$$

To obtain a solution, $\bar{v}(z, p)$, $f(z)$, $g(z)$ and $\bar{q}(z, p)$ can be expanded into a series of eigenfunctions of a beam of perfectly elastic material as

$$\bar{v}(z, p) = \sum_{n=1}^{\infty} F_n(p)v_n(z), \qquad f(z) = \sum_{n=1}^{\infty} f_n v_n(z),$$

$$g(z) = \sum_{n=1}^{\infty} g_n v_n(z), \qquad \bar{q}(z, p) = \sum_{n=1}^{\infty} P_n(p)v_n(z). \tag{4.45}$$

The functions $v_n(z)$ are orthogonal and normalized and satisfy the equation

$$\frac{d^4}{dz^4}v_n(z) - \lambda_n^4 v_n = 0, \tag{4.46}$$

where

$$\lambda_n^4 = \omega_n^2/\beta_0^2, \qquad \beta_0^2 = EI/(\rho_b A_b). \tag{4.47}$$

Substituting relations (4.45) in Eq. (4.38), we get

$$[\beta^2(p)\lambda_n^4 + p^2]F_n(p) = \frac{1}{\rho_b A_b}P_n(p) + pf_n + g_n. \tag{4.48}$$

Hence,

$$F_n(p) = \frac{[1/(\rho_b A_b)]P_n(p) + pf_n + g_n}{\beta^2(p)\lambda_n^4 + p^2}. \tag{4.49}$$

Therefore,

$$\bar{v}(z, p) = \sum_{n=1}^{\infty} \frac{[1/(\rho_b A_b)]P_n(p) + pf_n + g_n}{\beta^2(p)\lambda_n^4 + p^2}v_n(z), \tag{4.50}$$

$$\bar{v}(z, p) = \frac{1}{A_b}\sum_{n=1}^{\infty}\frac{v_n(z)\int_0^L \bar{q}(u, p)v_n(u)\,du}{\beta^2(p)\lambda_n^4 + p^2}$$

$$+ \sum_{n=1}^{\infty}\frac{v_n(z)\left[p\int_0^L f(u)v_n(u)\,du + \int_0^L g(u)v_n(u)\,du\right]}{\beta^2(p)\lambda_n^4 + p^2}. \tag{4.51}$$

The function $\bar{v}(z, p)$ can be obtained from Eq. (4.51), provided we know $\beta(p)$ which, in turn, depends on $\bar{E}(p)$, as defined in Eq. (4.43). $v(z, t)$ is then obtained by taking the inverse transform of (4.51).

4.3 ENERGY APPROACH

The differential equation approach discussed in Section 4.2 can be successfully applied for beams with a constant cross-section. For beams where the cross-section is variable, the mass and the flexural rigidity are complicated functions of the spatial coordinate. In such situations, the energy method which gives an upper bound to the exact solution is effective. The method assumes that we have a knowledge of the approximate mode shape that satisfies the geometric (or kinematic) boundary conditions.

4.3.1 RAYLEIGH'S METHOD

Rayleigh's method is apt for obtaining the fundamental frequency of a vibrating system. It is based on the principle of conservation of energy.

For a system undergoing simple harmonic motion, the maximum kinetic energy is equal to the maximum potential energy. The potential energy V per unit length in bending is

$$V = \tfrac{1}{2} \int_0^L EI(\frac{\partial^2 v}{\partial z^2})^2 \, dz.$$

The kinetic energy T per unit length for an element is

$$dT = \frac{m_b}{2}(\frac{\partial v}{\partial t})^2 \, dz. \tag{4.52}$$

Therefore,

$$T = \tfrac{1}{2} \int_0^L m_b(\frac{\partial v}{\partial t})^2 \, dz. \tag{4.53}$$

For the simple harmonic motion of a beam, we write

$$v(z, t) = v(z) \sin \omega t. \tag{4.54}$$

$$T_{\max} = \frac{\omega^2}{2} \int_0^L m_b v^2 \, dz, \tag{4.55}$$

$$V_{\max} = \tfrac{1}{2} \int_0^L EI(\frac{d^2 v}{dz^2})^2 \, dz. \tag{4.56}$$

From the conservation of energy, we then have

$$\omega^2 = \frac{\displaystyle\int_0^L EI(d^2v/dz^2)^2 \, dz}{\displaystyle\int_0^L m_b v^2 \, dz}. \tag{4.57}$$

Example 4.2   Consider a uniform cantilever beam whose deflection curve is given by

$$v(z) = v_0(1 - \cos\frac{\pi z}{2L}), \tag{4.58}$$

where v_0 is the maximum amplitude which exists at the tip. Substituting for $v(z)$ in Eq. (4.57), we have

$$\omega^2 = \frac{\displaystyle\int_0^L EI(\cos\frac{\pi z}{2L})^4(\frac{\pi}{2L})^4 \, dz}{\displaystyle\int_0^L m_b(1 - \cos\frac{\pi z}{2L})^2 \, dz}. \tag{4.59}$$

From this, we get

$$f = \frac{\omega}{2\pi} = \frac{34.93}{L^2}(\frac{EI}{m_b})^{1/2}, \tag{4.60}$$

$$f_{\text{exact}} = \frac{33.63}{L^2}(\frac{EI}{m_b})^{1/2}.$$

(4.61)

The error is approximately 4 per cent. This can be further reduced by choosing a better displacement function.

4.3.2 RAYLEIGH-RITZ METHOD

The Rayleigh-Ritz method is a modification of Rayleigh's approach. In this method, the displacement function is a combined function of several variables which are such that the boundary conditions are individually satisfied. The advantage gained thereby is that the series can be truncated, depending on the desired accuracy. For instance, let

$$v(z) = a_1 v_1(z) + a_2 v_2(z) + \ldots,$$

(4.62)

where $a_1, a_2, \ldots$ are undetermined parameters, so chosen as to minimize the frequency. The minimization results in

$$\frac{\partial}{\partial a_i} \left[\frac{\int_0^L EI(\frac{d^2v}{dz^2})^2 \, dz}{\int_0^L m_b v^2 \, dz} \right] = 0.$$

(4.63)

This reduces to

$$\left(\int_0^L m_b v^2 \, dz \right) \frac{\partial}{\partial a_i} \int_0^L EI(\frac{d^2v}{dz^2})^2 \, dz - \int_0^L EI(\frac{d^2v}{dz^2}) \, dz \frac{\partial}{\partial a_i} \int_0^L m_b v^2 \, dz = 0.$$

(4.64)

Combining Eqs. (4.57) and (4.64), we have

$$\frac{\partial R}{\partial a_i} = 0 \qquad (i = 1, 2, \ldots),$$

(4.65)

where

$$R = \int_0^L EI(\frac{d^2v}{dz^2})^2 \, dz - \omega^2 \int_0^L m_b v^2 \, dz.$$

Equations (4.65) result in linear, homogeneous, algebraic, and simultaneous equations in $a_1, a_2, \ldots$. For a nontrivial solution, the determinant of the coefficients must be equal to zero, leading to the characteristic equation.

Example 4.3   Consider the cantilever beam described in Example 4.2. But now let a two-term approximation

$$v(z) = a_1 z^2 + a_2 z^3$$

(4.66)

represent the displacement function. The two terms individually satisfy the geometric boundary conditions. Applying (4.65), we get

$$\frac{\partial R}{\partial a_1} \equiv (8a_1 L + 12a_2 L^2)EI - m_b \omega^2(\tfrac{2}{15} a_1 L^5 + \frac{a_2}{3}L^6) = 0,$$

$$\frac{\partial R}{\partial a_2} \equiv (24a_2 L + 12a_1 L^2)EI - m_b \omega^2(\tfrac{2}{7}a_2 L^7 + a_1 L^6) = 0.$$

(4.67)

The circular frequency ω_1 is

$$\omega_1 = \frac{3.53}{L^2}(\frac{EI}{m_b})^{1/2},$$

(4.68)

$$\omega_{1 \text{ (exact)}} = \frac{3.515}{L^2} \left(\frac{EI}{m_b}\right)^{1/2}. \tag{4.69}$$

In this approach, in addition to the first frequency, we get higher harmonics. However, the convergence to the higher modes is poor. This can be improved by taking a higher-order function for the displacement and by making use of beam characteristic functions.

4.4 COUPLED BENDING AND TORSIONAL VIBRATION OF BEAMS

In our discussion on transverse vibration, we have assumed that the beam vibrates in its plane of symmetry. If this is not so, then the lateral vibration is usually coupled with torsional vibration. We have also noticed (see Section 2.6) that, in the buckling of open sections, coupling exists whenever the shear centre does not coincide with the centre of gravity of the section.

When the cross-section of the beam has an axis of symmetry, one of the two bending vibrations is independent of the other vibrations (one bending and one torsional). This results in a coupling of the other bending vibration and the torsional vibration, as is observed in channel sections. The simplest case of the existence of torsional and bending vibrations arises when the shear centre coincides with the centre of gravity of the section, for example, for an *I*-beam with two axes of symmetry. Here, both the bending vibrations and the torsional vibration are independent of each other and the lowest frequency depends on the relative bending and torsional rigidities. Normally, the first natural frequency corresponds to the bending mode, the second to the torsional mode, and so on.

Let us now consider a beam with an arbitrary cross-section, as shown in Fig. 2.7. Here, the principal centroidal axes pass through the point O; x_0 and y_0 are the coordinates of the shear centre C. During vibration, the cross-section undergoes translation and rotation. Since the system is assumed to be linear, we can consider the two deformations independently and superimpose the solutions.

As stated in Section 2.6, the final deflection of the centroid comprises the deflection (due to translation), and the component of deflection (due to rotation). Hence, when the beam vibrates, the inertia forces of translation in the X- and Y-direction are

$$-\rho_b A_b \frac{\partial^2}{\partial t^2}(u + y_0\beta), \qquad -\rho_b A_b \frac{\partial^2}{\partial t^2}(v - x_0\beta). \tag{4.70}$$

These forces act through the centroid. The cross-section rotates about the shear centre C to which the inertia forces must be transferred. The net inertial moment is given by

$$-\rho_b A_b I_p \frac{\partial^2\beta}{\partial t^2} + A_b \frac{\partial^2}{\partial t^2}(v - x_0\beta)x_0 - \rho_b A_b \frac{\partial^2}{\partial t^2}(u + y_0\beta)y_0, \tag{4.71}$$

where I_p is the polar moment of inertia about the centroid.

Substituting expression (4.71) for the inertia forces, we obtain, instead of the force due to the load P in Eqs. (2.57)–(2.59), three simultaneous differential equations:

$$EI_y \frac{\partial^4 u}{\partial z^4} + \rho_b A_b \frac{\partial^2 u}{\partial t^2} + \rho_b A_b y_0 \frac{\partial^2\beta}{\partial t^2} = 0, \tag{4.72}$$

$$EI_x \frac{\partial^4 v}{\partial z^4} + \rho_b A_b \frac{\partial^2 v}{\partial t^2} - \rho_b A_b x_0 \frac{\partial_2\beta}{\partial t^2} = 0, \tag{4.73}$$

$$E\Gamma \frac{\partial^2\beta}{\partial z^4} - GJ \frac{\partial^2\beta}{\partial z^2} - \rho_b A_b x_0 \frac{\partial^2 v}{\partial t^2} + \rho_b A_b y_0 \frac{\partial^2 u}{\partial t^2} + \rho_b A_b I_0 \frac{\partial^2\beta}{\partial t^2} = 0, \tag{4.74}$$

where I_0 is the polar moment of inertia of the cross-section about the shear centre C.

If the beam has two axes of symmetry, then $x_0 = 0$, $y_0 = 0$ and Eqs. (4.72), (4.73), and (4.74) become independent of one another. The first two equations represent the uncoupled bending vibrations in the X- and Y-direction and the last equation gives the uncoupled torsional vibrations. If the X-axis is an axis of symmetry, implying that $y_0 = 0$, then (4.72) is independent of (4.73) and (4.74).

The procedure for solving (4.72)–(4.74) entails assuming a solution that satisfies the boundary conditions for the displacements u, v, and rotation β. This is possible when similar types of boundary conditions exist for all the three. For example, if the beam is simply-supported, its ends are free to warp and rotate about the X- and Y-axis but can neither rotate about the Z-axis nor deflect in the X- and Y-direction. In such a case, the boundary conditions are

$$u = v = \beta = 0 \qquad (z = 0, L),$$

$$\frac{\partial^2 u}{\partial z^2} = \frac{\partial^2 v}{\partial z^2} = \frac{\partial^2 \beta}{\partial z^2} = 0 \qquad (z = 0, L). \tag{4.75}$$

For boundary conditions of the form

$$u = v = \beta = 0 \qquad (z = 0, L),$$

$$\frac{\partial^2 u}{\partial z^2} = \frac{\partial^2 v}{\partial z^2} = \frac{\partial \beta}{\partial z} = 0 \qquad (z = 0, L), \tag{4.76}$$

where the ends of the beam are free to rotate about the X- and Y-axis but are restrained from warping, it is not possible to obtain a solution by employing the differential equation approach. Normally, in such cases, the Rayleigh-Ritz method, based on the principle of conservation of the total energy of the system, helps determine the approximate values of the frequencies of vibrations.

4.5 NUMERICAL TECHNIQUES

The techniques discussed in Sections 4.2–4.4 are suitable for simple problems. However, if a computer is to be used for the analysis, it is best to employ a technique well suited for computer applications.

4.5.1 FINITE DIFFERENCE TECHNIQUE

In Section 2.7.1, the finite difference technique has been applied to obtain the buckling load of a column. The very same formulation may be applied to the vibration of beams, but with a slight modification: the second term of Eq. (2.8) is replaced by inertia forces. (The inertial force is a product of mass per unit length and the second derivative of displacement with respect to time t.) The governing equation may then be written as

$$EI \frac{\partial^4 v}{\partial z^4} + \rho_b A_b \frac{\partial^2 v}{\partial t^2} = 0. \tag{4.77}$$

This partial differential equation can be solved by converting it into an ordinary differential equation. This is possible if we assume the motion to be simple harmonic. In most problems where we seek natural frequency, the finite difference technique can be used.

Equation (4.77) can be solved also by assuming finite differences separately for time and for space. Let

$$v(z, t) = v(z)e^{i\omega t} \tag{4.78}$$

Substituting relation (4.78) in (4.77), we get

$$d^4v/dz^4 - \lambda_0^2 v = 0, \tag{4.79}$$

where $\lambda_0^2 = \rho_b A_b \omega^2 / EI$ is defined as the frequency parameter for a uniform beam.

Example 4.4   Consider a beam which is simply-supported at the right end ($z = L$), clamped at the left end ($z = 0$), and undergoing free vibrations (see Fig. 4.3). The boundary conditions are

$$v(0) = \frac{dv}{dz}(0) = v(L) = \frac{d^2v}{dz^2}(L) = 0. \tag{4.80}$$

The finite difference problem is obtained by substituting the central difference operators in the

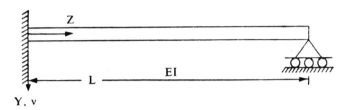

Fig. 4.3   Clamped-Hinged Uniform Beam.

differential equation and the boundary conditions, as done for the derivatives in Section 2.7.1. Introducing the nondimensional parameter ξ ($=z/L$), we have

$$d^4v/d\xi^4 - \lambda^2 v = 0, \tag{4.81}$$

where

$$\lambda^2 = \lambda_0^2 L^4. \tag{4.82}$$

Dividing the interval (0, 1) of ξ into n equal parts of length $h = 1/n$ and multiplying throughout by h^4, Eq. (4.81) becomes

$$h^4 \frac{d^4v}{d\xi^4} - \frac{\lambda^2}{n^4} v = 0. \tag{4.83}$$

Substituting relation (2.68) for the fourth derivative at point i, we get

$$v_{i+2} - 4v_{i+1} + (6 - \lambda^2/n^4)v_i - 4v_{i-1} + v_{i-2} = 0. \tag{4.84}$$

The boundary conditions (4.80) are transformed into central difference conditions as

$$v_0 = 0, \qquad v_{-1} = v_1, \qquad v_n = 0, \qquad v_{n+1} = -v_{n-1}. \tag{4.85}$$

Equation (4.84) is applicable only to internal pivot points. The difference problem is then solved by assigning different values to n, and the accuracy of the solution is increased by raising the value of n in stages. For instance, for $n = 2$, we obtain

$$v_3 - 4v_2 + (6 - \lambda^2/2^4)v_1 - 4v_0 + v_{-1} = 0. \tag{4.86}$$

Using the boundary conditions, we have

$$(6 - \lambda^2/2^4)v_1 = 0. \tag{4.87}$$

For a nontrivial solution, v_1 is not equal to zero. Hence,

$$\lambda = 4\sqrt{6}. \tag{4.88}$$

Therefore,

$$\omega = \frac{9.8}{L^2}(\frac{EI}{\rho_b A_b})^{1/2}. \tag{4.89}$$

For $n = 3$, we get

$$v_3 - 4v_2 + (6 - \lambda^2/3^4)v_1 - 4v_0 + v_{-1} = 0,$$
$$v_4 - 4v_3 + (6 - \lambda^2/3^4)v_2 - 4v_1 + v_0 = 0. \tag{4.90}$$

After substituting the boundary conditions, Eqs. (4.90) reduce to

$$(7 - \lambda^2/3^4)v_1 - 4v_2 = 0,$$
$$-4v_1 + (5 - \lambda^2/3^4)v_2 = 0. \tag{4.91}$$

For a nontrivial solution, the determinant of the coefficients of v_1 and v_2 must be zero. This leads to the characteristic equation

$$\mu^2 - 12\mu + 19 = 0, \tag{4.92}$$

where $\mu = \lambda^2/3^4$. The two roots of Eq. (4.92) are $\mu_1 = 1.94$, $\mu_2 = 10.06$. Therefore,

$$\omega_1 = \frac{12.60}{L^2}(\frac{EI}{\rho_b A_b})^{1/2}. \tag{4.93}$$

By assigning still higher values to n, it is possible to ensure convergence of the frequency to the exact solution.

4.5.2   FINITE ELEMENT TECHNIQUE

The finite element method is based on the fact that a continuum can be represented by a group of discrete elements, each element being essentially a part of the whole and satisfying the equation of equilibrium and of compatibility. In Sections 2.7.2 and 3.6.2 we used the principle of minimum potential energy to obtain the elastic and geometric stiffness matrices of the structure. Here, we employ the variational principle to derive the finite element equation.

The variational principle best suited for problems in dynamics is Hamilton's principle. The functional for this variational principle, as in the principle of minimum potential energy and the principle of complementary energy, is the Lagrangian L which is defined as

$$L = \text{kinetic energy } (T) - \text{strain energy } (U) - \text{potential of applied external forces.}$$

Hamilton's principle is given by

$$\int_{t_1}^{t_2} L \, dt = 0. \tag{4.94}$$

This is expressed as: *Among all possible time histories of displacement configurations, which satisfy compatibility and geometric boundary conditions and which also satisfy the conditions at times t_1 and t_2, the history that is the actual solution makes the Lagrangian functional a minimum.*

In dynamic problems, the displacements, velocities, stresses, strains, and loadings are all functions of time. For an interpolation type displacement model, we have

$$\{u(x, y, z, t)\} = [N(x, y, z)]\{q(t)\}, \tag{4.95}$$

where $\{u\}$ is the displacement vector, $[N]$ the matrix of shape functions, and $\{q\}$ the vector of nodal displacements. From Eqs. (2.89) and (2.92), we know

$$\{e\} = [B]\{q\},$$

$$\{\sigma\} = [\chi]\{e\} = [\chi][B]\{q\}.$$

The Lagrangian L may then be expressed as

$$L = \tfrac{1}{2} \int_V (\rho_b \dot{u}^T \dot{u} - e^T[\chi]e + 2u^T X)\, dv + \int_{s_1} u^T \varphi\, ds_1, \tag{4.96}$$

where X is the vector of body forces per unit volume, φ the vector of surface tractions, and s_1 the part of the surface over which surface tractions are prescribed.

Substituting Eqs. (4.95), (2.89), and (2.92) in Eq. (4.96), we get

$$-L = \tfrac{1}{2} \int_V (q^T[B]^T[\chi][B]q - \rho_b \dot{q}^T[N]^T[N]q - 2q^T[N]^T X)\, dv - \int_{s_1} q^T[N]^T \varphi\, ds_1. \tag{4.97}$$

Integrating Eq. (4.97) and taking the variation, we obtain

$$\int_{t_1}^{t_2} \left(\delta q^T \int_V [B]^T[\chi][B]\, dv\, q - \delta \dot{q}^T \int_V [N]^T[N]\rho_b \dot{q}\, dv - \delta q^T \int_V [N]^T X\, dv \right.$$
$$\left. - \delta q^T \int_{s_1} [N]^T \varphi\, ds_1 \right) dt = 0. \tag{4.98}$$

By integrating the second term in parts with respect to time t, we obtain

$$\int_{t_1}^{t_2} \delta \dot{q}^T \int_V \rho_b[N]^T \dot{q}[N]\, dv\, dt = \left(\delta q^T \int_V \rho_b[N]^T[N]\dot{q}\, dv \right)_{t_1}^{t_2}$$
$$- \int_{t_1}^{t_2} \delta q^T \int_V \rho_b[N]^T[N]\ddot{q}\, dv\, dt. \tag{4.99}$$

As per Hamilton's principle, the tentative displacement configuration must satisfy the given conditions at time t_1 and t_2. Hence $\delta q(t_1) = \delta q(t_2) = 0$. In view of this, the first term on the right-hand side of Eq. (4.99) vanishes. Equation (4.98) may then be rewritten as

$$\int_{t_1}^{t_2} \delta q^T \left(\int_V \rho_b[N]^T[N]\, dv\, \ddot{q} + \int_V [B]^T[\chi][B]q\, dv \right.$$
$$\left. - \int_V [N]^T X\, dv - \int_{s_1} [N]^T \varphi\, ds_1 \right) dt = 0. \tag{4.100}$$

As δq is arbitrary, the quantity within the parentheses must be equal to zero in order that Eq. (4.100) is satisfied. This gives the equation of motion

$$[M]\{\ddot{q}\} + [K]\{q\} = \{Q\}, \tag{4.101}$$

where $[K]$ is the stiffness matrix, $\{Q\}$ the load vector, and $[M]$ the consistent mass matrix which may be expressed as

$$[M] = \rho_b \int_V [N]^T[N]\, dv.$$

The mass matrix is termed "consistent" because the same displacement function is used for obtaining the elastic stiffness matrix and the mass matrix. The major advantage of the consistent formulation is that it enables us to get accurate mode shapes and frequencies which are upper bound to the exact solution.

When the mass is concentrated at the nodes, the formulation is said to be "lumped".

The lumped mass matrix has several advantages because it is diagonal: it occupies less storage space than the consistent mass matrix; it offers facility for matrix calculations; it enables both rotary and translation inertia to be treated in the same way. The lumped mass formulation is therefore the natural choice in many practical problems.

We have already discussed the method of obtaining the element stiffness matrix (see Section 2.7.2). Therefore, we shall now confine our study to the technique of obtaining the consistent mass matrix.

Consistent Mass Matrix for Pin-Jointed Bar Element

Let $\overline{X}$, $\overline{Y}$, $\overline{Z}$ represent the local coordinate system, and X, Y, Z the global coordinate system, as shown in Fig. 4.4. Here, U_1, U_2, . . . , U_6 represent the global displacements at nodes 1 and 2.

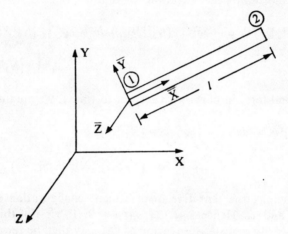

Fig. 4.4   Pin-Jointed Bar Element.

The displacements at a point x from node 1 are given by

$$u_x = U_1 + (U_4 - U_1)\xi,$$
$$u_y = U_2 + (U_5 - U_2)\xi,$$
$$u_z = U_3 + (U_6 - U_3)\xi,$$

(4.102)

where $\xi = (x/l)$. Equations (4.102) may be written in the matrix form as

$$\{u\} = [N]\{U\},$$

$$\begin{Bmatrix} u_x \\ u_y \\ u_z \end{Bmatrix} = \begin{bmatrix} 1-\xi & 0 & 0 & \xi & 0 & 0 \\ 0 & 1-\xi & 0 & 0 & \xi & 0 \\ 0 & 0 & 1-\xi & 0 & 0 & \xi \end{bmatrix} \begin{bmatrix} U_1 \\ U_2 \\ \vdots \\ U_6 \end{bmatrix}.$$

(4.103)

If ρ_b is the density and A_b the cross-section area, then we have

$$[m] = \int_v \rho_b [N]^T [N] \, dv = \rho_b A_b l \int_0^1 [S(\xi)] \, d\xi,$$

(4.104)

where

$$[S(\xi)] = \begin{bmatrix} (1-\xi)^2 & 0 & 0 & (1-\xi)\xi & 0 & 0 \\ 0 & (1-\xi)^2 & 0 & 0 & \xi(1-\xi) & 0 \\ 0 & 0 & (1-\xi)^2 & 0 & 0 & \xi(1-\xi) \\ (1-\xi)\xi & 0 & 0 & \xi^2 & 0 & 0 \\ 0 & (1-\xi)\xi & 0 & 0 & \xi^2 & 0 \\ 0 & 0 & (1-\xi)\xi & 0 & 0 & \xi^2 \end{bmatrix}.$$

Integration with respect to ξ is carried out for obtaining the mass matrix

$$[m] = \frac{\rho_b A_b l}{6} \begin{bmatrix} 2 & 0 & 0 & 1 & 0 & 0 \\ 0 & 2 & 0 & 0 & 1 & 0 \\ 0 & 0 & 2 & 0 & 0 & 1 \\ 1 & 0 & 0 & 2 & 0 & 0 \\ 0 & 1 & 0 & 0 & 2 & 0 \\ 0 & 0 & 1 & 0 & 0 & 2 \end{bmatrix}. \tag{4.105}$$

The mass matrix thus obtained is in the global coordinate system.

In many situations where it may not be possible to integrate the expressions analytically, numerical integration is used. It is obvious from Eq. (4.105) that some of the off diagonal elements are nonzero, thereby increasing the band width.

Mass Matrix in Local Coordinates

Since the bar element can have deformation along its axis alone, only one degree of freedom is possible at each node. Hence,

$$u_{\bar{x}} = [(1-\xi)\xi] \begin{Bmatrix} \bar{U}_1 \\ \bar{U}_2 \end{Bmatrix}. \tag{4.106}$$

Therefore,

$$[N] = [(1-\xi)\xi], \tag{4.107}$$

$$[m]_{\text{local}} = \int_v \rho_b [N]^T [N] \, dv = \frac{\rho_b A_b l}{6} \begin{bmatrix} 2 & 1 \\ 1 & 2 \end{bmatrix}. \tag{4.108}$$

The mass matrices derived here can be used in obtaining the natural frequency of the axial vibration of bars. Let us now consider a uniform bar which is fixed at one end and free at the other end, and discretized into two elements, as shown in Fig. 4.5. The nodal degrees of

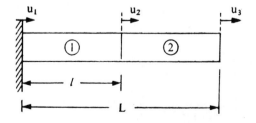

Fig. 4.5 Uniform Axial Bar.

freedom are u_1, u_2, and u_3. A linear displacement function of the type

$$u_x = \alpha_1 \xi + \alpha_2 \tag{4.109}$$

is chosen to represent the displacement at any point within the element. The elastic stiffness matrix and the consistent mass matrix can be obtained by applying the procedure just described. The elastic stiffness matrices of elements 1 and 2 are

$$[k^{(1)}] = \frac{A_b E}{l}\begin{matrix} u_1 & u_2 \\ \begin{bmatrix} 1 & -1 \\ -1 & 1 \end{bmatrix} \end{matrix} = \frac{2A_b E}{L}\begin{matrix} u_1 & u_2 \\ \begin{bmatrix} 1 & -1 \\ -1 & 1 \end{bmatrix} \end{matrix}, \tag{4.110}$$

$$[k^{(2)}] = \frac{A_b E}{l}\begin{matrix} u_2 & u_3 \\ \begin{bmatrix} 1 & -1 \\ -1 & 1 \end{bmatrix} \end{matrix} = \frac{2A_b E}{L}\begin{matrix} u_2 & u_3 \\ \begin{bmatrix} 1 & -1 \\ -1 & 1 \end{bmatrix} \end{matrix}. \tag{4.111}$$

To obtain the overall stiffness matrix, each element stiffness matrix is enlarged so as to include all the degrees of freedom and then these matrices are added by applying the matrix algebra. That is,

$$[k^{(1)}] = \frac{2A_b E}{L}\begin{matrix} u_1 & u_2 & u_3 \\ \begin{bmatrix} 1 & -1 & 0 \\ -1 & 1 & 0 \\ 0 & 0 & 0 \end{bmatrix} \end{matrix}, \tag{4.112}$$

$$[k^{(2)}] = \frac{2A_b E}{L}\begin{matrix} u_1 & u_2 & u_3 \\ \begin{bmatrix} 0 & 0 & 0 \\ 0 & 1 & -1 \\ 0 & -1 & 1 \end{bmatrix} \end{matrix}, \tag{4.113}$$

$$[K] = [k^{(1)}] + [k^{(2)}] = \frac{2A_b E}{L}\begin{matrix} u_1 & u_2 & u_3 \\ \begin{bmatrix} 1 & -1 & 0 \\ -1 & 2 & -1 \\ 0 & -1 & 1 \end{bmatrix} \end{matrix}. \tag{4.114}$$

From the boundary conditions of the problem, we get $u_1 = 0$. Hence,

$$[K] = \frac{2A_b E}{L}\begin{matrix} u_2 & u_3 \\ \begin{bmatrix} 2 & -1 \\ -1 & 1 \end{bmatrix} \end{matrix}. \tag{4.115}$$

Similarly, we obtain the consistent mass matrices for elements 1 and 2 as

$$[m^{(1)}] = \frac{\rho_b A_b l}{6}\begin{matrix} u_1 & u_2 \\ \begin{bmatrix} 2 & 1 \\ 1 & 2 \end{bmatrix} \end{matrix} = \frac{\rho_b A_b L}{12}\begin{matrix} u_1 & u_2 \\ \begin{bmatrix} 2 & 1 \\ 1 & 2 \end{bmatrix} \end{matrix}, \tag{4.116}$$

$$[m^{(2)}] = \frac{\rho_b A_b l}{6}\begin{matrix} u_2 & u_3 \\ \begin{bmatrix} 2 & 1 \\ 1 & 2 \end{bmatrix} \end{matrix} = \frac{\rho_b A_b L}{12}\begin{matrix} u_2 & u_3 \\ \begin{bmatrix} 2 & 1 \\ 1 & 2 \end{bmatrix} \end{matrix}. \tag{4.117}$$

Therefore, the overall mass matrix is

$$[M] = [m^{(1)}] + [m^{(2)}] = \frac{\rho_b A_b L}{12} \begin{matrix} u_1 & u_2 & u_3 \\ \begin{bmatrix} 2 & 1 & 0 \\ 1 & 4 & 1 \\ 0 & 1 & 2 \end{bmatrix} \end{matrix}. \tag{4.118}$$

Applying the constraint on u_1, we have

$$[M] = \frac{\rho_b A_b L}{12} \begin{matrix} u_2 & u_3 \\ \begin{bmatrix} 4 & 1 \\ 1 & 2 \end{bmatrix} \end{matrix}. \tag{4.119}$$

For the free vibration problem, the right-hand side of Eq. (4.101) is zero; and for the problem under consideration, $\{q\}$ corresponds to $\{u\}$. Hence,

$$[M]\{\ddot{u}\} + [K]\{u\} = 0. \tag{4.120}$$

Assuming the motion to be simple harmonic, we get

$$u(z, t) = u(z)e^{i\omega t}.$$

After substituting for $[M]$ and $[K]$, Eq. (4.120) may be rewritten as

$$\left[-\omega^2 \frac{\rho_b A_b L}{12} \begin{bmatrix} 4 & 1 \\ 1 & 2 \end{bmatrix} + \frac{2A_b E}{L} \begin{bmatrix} 2 & -1 \\ -1 & 1 \end{bmatrix} \right]\{u\} = 0. \tag{4.121}$$

The condition for the nontrivial solution for $\{u\}$ is that the determinant formed by the coefficient in Eq. (4.121) must be equal to zero. Defining $\lambda^2 = \omega^2 \rho_b L^2/(24E)$, we have

$$\begin{vmatrix} 2(1 - 2\lambda^2) & -(1 + \lambda^2) \\ -(1 + \lambda^2) & (1 - 2\lambda^2) \end{vmatrix} = 0. \tag{4.122}$$

Expanding the determinant, we get

$$7\lambda^4 - 10\lambda^2 + 1 = 0. \tag{4.123}$$

The roots of Eq. (4.123) are $\lambda_1^2 = (5 - 3\sqrt{2})/7$, $\lambda_2^2 = (5 + 3\sqrt{2})/7$. Hence,

$$\omega_1 = 1.6114\sqrt{E/\rho_b L^2}, \qquad \omega_2 = 5.6293\sqrt{E/\rho_b L^2}. \tag{4.124}$$

These values are 2.6 per cent and 19.5 per cent higher than the exact values for the first and second modes, respectively.

Once the stiffness and mass matrices are assembled, the eigenvalues and eigenvectors may be obtained by different methods. Some of the well known methods are (a) power method, (b) Jacobi method, and (c) subspace iteration method. The choice of method depends on the size of the problem and the number of eigenvalues desired. For the subspace iteration technique, the time required for one eigenvalue remains almost the same as that for several eigenvalues.

Let us now study a slightly more complicated problem, namely, the transverse vibration of beams. Consider a uniform cantilever beam of length L which is clamped at one end and free at the other end (see Fig. 4.6). If we are interested only in the first two modes, we may neglect the effects of shear deformation and rotatory inertia. As there are four nodal degrees of freedom for each element, the displacement function must have a minimum of four constants. Let the function be given by

$$v(z) = \alpha_1 + \alpha_2 z + \alpha_3 z^2 + \alpha_4 z^3. \tag{4.125}$$

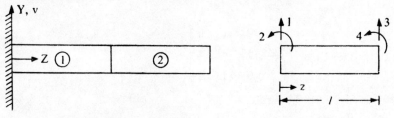

(a) Uniform cantilever beam   (b) Finite element idealization

Fig. 4.6   Uniform Cantilever Beam.

The constants $\alpha_1, \ldots, \alpha_4$ are determined by making use of

$$v_1 = v\Big|_{z=0}, \qquad v_2 = \frac{dv}{dz}\Big|_{z=0}, \qquad v_3 = v\Big|_{z=l}, \qquad v_4 = \frac{dv}{dz}\Big|_{z=l}. \tag{4.126}$$

Equation (4.125) may now be written in terms of the nondimensional parameter ξ as

$$v = \begin{bmatrix} (1 - 3\xi^2 + 2\xi^3) \\ (\xi - 2\xi^2 + \xi^3)l \\ (3\xi^2 - 2\xi^3) \\ (-\xi^2 + \xi^3)l \end{bmatrix}^T \begin{Bmatrix} v_1 \\ v_2 \\ v_3 \\ v_4 \end{Bmatrix}$$

$$= [N]\{v\}, \tag{4.127}$$

$$\epsilon_{zz} = -\frac{1}{l^2}\, y\, \frac{\partial^2 v}{\partial \xi^2}$$

$$= \frac{-y^2}{l^2} \begin{bmatrix} 6(-1 + 2\xi) \\ 2(-2 + 3\xi)l \\ 6(1 - 2\xi) \\ 2(-1 + 3\xi)l \end{bmatrix}^T \begin{Bmatrix} v_1 \\ v_2 \\ v_3 \\ v_4 \end{Bmatrix}$$

$$= [B]\{v\}. \tag{4.128}$$

If we know $[B]$, we can obtain the element stiffness matrices $[k]$ as

$$[k^{(1)}] = \frac{EI}{l^3} \begin{matrix} \quad\; v_1 \quad\;\; v_2 \quad\;\; v_3 \quad\;\; v_4 \\ \begin{bmatrix} 12 & 6l & -12 & 6l \\ 6l & 4l^2 & -6l & 2l^2 \\ -12 & -6l & 12 & -6l \\ 6l & 12l^2 & -6l & 4l^2 \end{bmatrix} \end{matrix}, \tag{4.129}$$

$$[k^{(2)}] = \frac{EI}{l^3} \begin{matrix} \quad\; v_3 \quad\;\; v_4 \quad\;\; v_5 \quad\;\; v_6 \\ \begin{bmatrix} 12 & 6l & -12 & 6l \\ 6l & 4l^2 & -6l & 2l^2 \\ -12 & -6l & 12 & -6l \\ 6l & 12l^2 & -6l & 4l^2 \end{bmatrix} \end{matrix} \tag{4.130}$$

Assembling the matrices $[k]$ for the entire beam and applying the boundary conditions ($v_1 = 0$, $v_2 = 0$), we have

$$[K] = \frac{EI}{l^3} \begin{bmatrix} \overset{v_3}{24} & \overset{v_4}{0} & \overset{v_5}{-12} & \overset{v_6}{6l} \\ 0 & 8l^2 & -6l & 2l^2 \\ -12 & -6l & 12 & -6l \\ -6l & 2l^2 & -6l & 4l^2 \end{bmatrix}. \tag{4.131}$$

Using the matrix $[N]$ given by Eq. (4.127), we get the consistent element mass matrices as

$$[m^{(1)}] = \frac{\rho_b A_b l}{420} \begin{bmatrix} \overset{v_1}{156} & \overset{v_2}{22l} & \overset{v_3}{54} & \overset{v_4}{-13l} \\ 22l & 4l^2 & 13l & -3l^2 \\ 54 & 13l & 156 & -22l \\ -13l & -3l^2 & -22l & 4l^2 \end{bmatrix}, \tag{4.132}$$

$$[m^{(2)}] = \frac{\rho_b A_b l}{420} \begin{bmatrix} \overset{v_3}{156} & \overset{v_4}{22l} & \overset{v_5}{54} & \overset{v_6}{-13l} \\ 22l & 4l^2 & 13l & -3l^2 \\ 54 & 13l & 156 & -22l \\ -13l & -3l^2 & -22l & 4l^2 \end{bmatrix}, \tag{4.133}$$

$$[M] = \frac{\rho_b A_b l}{420} \begin{bmatrix} \overset{v_3}{312} & \overset{v_4}{0} & \overset{v_5}{54} & \overset{v_6}{-13l} \\ 0 & 3l^2 & 13l & -3l^2 \\ 54 & 13l & 156 & -22l \\ -13l & -3l^2 & -22l^2 & 4l^2 \end{bmatrix}. \tag{4.134}$$

Substituting the expressions for $[K]$ and $[M]$ in (4.120), we finally obtain

$$\left[-\omega^2 \frac{\rho_b A_b l}{420} \begin{bmatrix} 312 & 0 & 54 & -13l \\ 0 & 8l^2 & 13l & -3l^2 \\ 54 & 13l & 156 & -22l \\ -13l & -3l^2 & -22l & 4l^2 \end{bmatrix} + \frac{EI}{l^3} \begin{bmatrix} 24 & 0 & -12 & 6l \\ 0 & 8l^2 & -6l & 2l^2 \\ -12 & -6l & 12 & -6l \\ 6l & 2l^2 & -6l & 4l^2 \end{bmatrix} \right] v = 0. \tag{4.135}$$

We can now derive the eigenvalues from Eq. (4.135).

BIBLIOGRAPHY

Hurty, W. C. and Rubinstein, M. F., Dynamics of Structures, Prentice-Hall, Englewood Cliffs, New Jersey, 1964.

Martin, H. C., Introduction to Matrix Methods of Structural Analysis, McGraw-Hill, New York, 1966.

Martin, H. C. and Carey, G. F., Introduction to Finite Element Analysis, McGraw-Hill, New York, 1973.

Meirovitch, L., Analytical Methods in Vibration, Macmillan, New York, 1967.

Meirovitch, L., Elements of Vibration Analysis, McGraw-Hill, New York, 1975.

Nowacki, W., Dynamics of Elastic Systems, Chapman and Hall, London, 1963.

Przeminiecki, J. S., Theory of Matrix Structural Analysis, McGraw-Hill, New York, 1968.

Ralston, A. and Wilf, H. S., Mathematical Methods for Digital Computers, Wiley, New York, 1965.

Thomson, W. T., Theory of Vibrations with Applications, Prentice-Hall, Englewood Cliffs, New Jersey, 1972.

Timoshenko, S. P., Young, D. H., and Weaver, W., Jr., Vibration Problems in Engineering, 4th edn., Wiley, New York, 1974.

Zienkiewicz, O. C., The Finite Element Method in Engineering Science, 2nd edn., McGraw-Hill, London, 1971.

5

Vibration of Plates

5.1 INTRODUCTION

As stated in Chapter 3, plates are plane surface structures whose thickness is smaller than their length and width. They are bound either by straight lines or by curved lines and their boundary conditions may be free, simply-supported, or clamped edges, as well as elastic supports and elastic restraints. We restrict our study here to linear elastic materials which are homogeneous and isotropic.

The two-dimensional structural action of plates serves to lighten the structure and therefore offers numerous economic advantages. When a body is subjected to external forces, it deforms and produces internal stresses and strains. The dynamic loads carried by plates are predominantly normal to the plate surface. To understand the dynamic response of the structure, we shall introduce mathematical models. These models, as already pointed out, may be continuous or discrete and should be so chosen as to represent the real structure. The behaviour of continuous systems, like that of plates, is described by partial differential equations. A dynamic problem may be treated as a static problem by choosing the appropriate inertia forces.

5.2 GOVERNING EQUATION FOR TRANSVERSE MOTION

In Section 3.2, we considered the governing equation for the buckling of plates. For small oscillations, no inplane forces are developed, and so the inplane equilibrium equation is identically satisfied. The governing equation for the buckling analysis of plates is given by (3.18). However, for a free vibration problem, transverse load due to inplane forces is replaced by the inertia force which is associated with the lateral translation of a plate element and expressed as

$$p(x, y, t) = -m_p \frac{\partial^2 w}{\partial t^2}, \tag{5.1}$$

where $m_p (= \rho_p h)$ represents the mass of the plate per unit area.

In the dynamic analysis of plates, the lateral loads, and therefore the resultant deflections, are dependent on time. The differential equation for the forced, undamped motion of plates is

$$D\nabla^4 w = p_z(x, y, t) - m_p \frac{\partial^2 w}{\partial t^2}, \tag{5.2}$$

where x and y are the cartesian coordinates in the plane of the middle surface, as described in Section 4.2. For free vibration of plates, the external force p_z is zero, and the differential equation for the undamped motion becomes

$$D\nabla^4 w + m_p \frac{\partial^2 w}{\partial t^2} = 0. \tag{5.3}$$

Assuming a simple harmonic motion, we may write

$$w(x, y, t) = W(x, y)e^{i\omega t}, \tag{5.4}$$

where $W(x, y)$ is the shape function describing the modes of vibration and ω the natural frequency. Substituting Eq. (5.4) in Eq. (5.3), we get

$$\nabla^4 W(x, y) - \lambda^4 W(x, y) = 0, \tag{5.5}$$

where

$$\lambda^4 = m_p \omega^2 / D. \tag{5.6}$$

If the plate is subjected to inplane loadings, the differential equation for transverse motion becomes

$$D\nabla^4 w = N_x \frac{\partial^2 w}{\partial x^2} + N_y \frac{\partial^2 w}{\partial y^2} + 2N_{xy} \frac{\partial^2 w}{\partial x \, \partial y} - m_p \frac{\partial^2 w}{\partial t^2}. \tag{5.7}$$

5.2.1   FREE VIBRATION OF ISOTROPIC RECTANGULAR PLATES

Since the differential equation is linear in x, y, and t, we can adopt a variable-separable type of solution. Let

$$w(x, y, t) = W(x, y)e^{i\omega t} \tag{5.8}$$

and, further, let $W(x, y) = X_0(x) Y_0(y)$ which corresponds to certain types of boundary conditions. Substituting Eq. (5.8) in Eq. (5.5), we have

$$X_0^{IV}(x)Y_0(y) + 2X_0^{II}(x)Y_0^{II}(y) + X_0(x)Y_0^{IV}(y) - \lambda^4 X_0(x)Y_0(y) = 0. \tag{5.9}$$

Example 5.1   For simply-supported boundary conditions, the shape function may be represented by a double Fourier series, i.e.,

$$W(x, y) = X_0(x)Y_0(y) = \sum_{m=1}^{\infty} \sum_{n=1}^{\infty} W_{mn} \sin \frac{m\pi x}{a} \sin \frac{n\pi y}{b} \qquad (m, n = 1, 2, 3, \dots). \tag{5.10}$$

This equation satisfies both geometric and natural boundary conditions and its substitution in Eq. (5.9) gives

$$\frac{m^4 \pi^4}{a^4} + 2 \frac{m^2 \pi^2}{a^2} \frac{n^2 \pi^2}{b^2} + \frac{n^4 \pi^4}{b^4} - \lambda^4 = 0. \tag{5.11}$$

Hence,

$$\lambda^2 = \omega_{mn} \left(\frac{m_p}{D}\right)^{1/2} = \pi^2 \left(\frac{m^2}{a^2} + \frac{n^2}{b^2}\right), \tag{5.12}$$

$$\omega_{mn} = \pi^2 \left(\frac{m^2}{a^2} + \frac{n^2}{b^2}\right)\left(\frac{D}{m_p}\right)^{1/2}. \tag{5.13}$$

The fundamental mode of flexural vibration is a single sine wave in the X- and Y-direction and the natural frequency is associated with the values $m = 1$ and $n = 1$. When $m = 2$ and $n = 1$ or $n = 2$ and $m = 1$, the next two higher modes are obtained. These two modes have the same frequency, but the shapes associated with them are different. If this is so, the two vibration modes can be superimposed (see Fig. 5.1) in any ratio of their maximum displacements, provided the ratio a/b is rational. For a simply-supported square plate, we may express

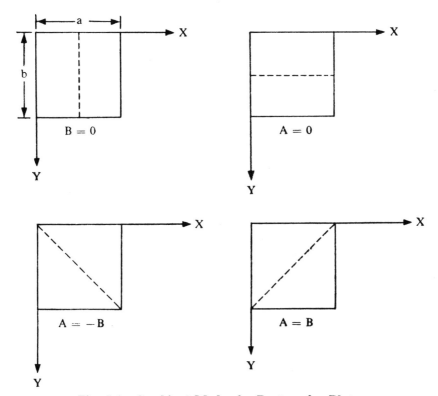

Fig. 5.1   Combined Modes for Rectangular Plates.

the two modes as

$$W(x, y) = A \sin \frac{2\pi x}{a} \sin \frac{\pi y}{b} + B \sin \frac{\pi x}{a} \sin \frac{2\pi y}{b}, \tag{5.14}$$

where A and B are arbitrary constants.

In many problems, the boundary conditions hinder the choice of shape functions along both the X- and Y-direction. If two of the opposite edges are simply-supported, then a sine function can be chosen along those edges and a general function along the remaining edges. This results in the conversion of the partial differential equation into an ordinary differential equation. Such conversion is referred to as the *levy type of solution*.

Example 5.2   Consider a rectangular plate which is simply-supported along the edges $x = 0$ and $x = a$. The shape function which satisfies the boundary conditions along the edges takes the form

$$W(x, y) = \sum_{m=1}^{\infty} Y_0(y) \sin \frac{m\pi x}{a}. \tag{5.15}$$

Substituting Eq. (5.15) with a specific value of m in (5.5) to determine $Y_0(y)$, we get the ordinary differential equation

$$\frac{d^4 Y_0}{dy^4} - \frac{2m^2\pi^2}{a^2} \frac{d^2 Y_0}{dy^2} + \left(\frac{m^4\pi^4}{a^4} - \frac{m_p\omega^2}{D}\right) Y_0 = 0. \tag{5.16}$$

The general solution for $Y_0(y)$ is

$$Y_0(y) = C_1 \cosh \alpha y + C_2 \sinh \alpha y + C_3 \cos \beta y + C_4 \sin \beta y, \qquad (5.17)$$

where

$$\alpha = [\omega(\frac{m_p}{D} + \frac{m^2\pi^2}{a^2})^{1/2}]^{1/2},$$

$$\beta = [\omega(\frac{m_p}{D} - \frac{m^2\pi^2}{a^2})^{1/2}]^{1/2}. \qquad (5.18)$$

The constants C_1, C_2, C_3, and C_4 are determined from the constraints along the edges $y = 0$ and $y = b$. These yield the frequency equation from which ω is determined. For instance, for a clamped edge, we have

$$(Y_0)_{y=0} = C_1 + C_3 = 0,$$

$$(\frac{dY_0}{dy})_{y=0} = \alpha C_2 + \beta C_4 = 0,$$

$$(Y_0)_{y=b} = C_1 \cosh \alpha b + C_2 \sinh \alpha b + C_3 \cos \beta b + C_4 \sin \beta b = 0,$$

$$(\frac{dY_0}{dy})_{y=b} = \alpha C_1 \sinh \alpha b + \alpha C_2 \cosh \alpha b - \beta C_3 \sin \beta b + \beta C_4 \cos \beta b = 0, \qquad (5.19)$$

which can be rewritten as

$$C_3(\alpha \cos \beta b - \alpha \cosh \alpha b) + C_4(\alpha \sin \beta b - \beta \sinh \alpha b) = 0,$$

$$C_3(-\beta \sin \beta b - \alpha \sinh \alpha b) + C_4(\beta \cos \beta b - \beta \cosh \alpha b) = 0. \qquad (5.20)$$

Equations (5.20) are homogeneous, algebraic equations, and hence, for a nontrivial solution, the determinant of the coefficients of C_3 and C_4 must be zero. This leads to

$$2\alpha\beta - 2\alpha\beta \cosh \alpha b \cos \beta b + (\alpha^2 - \beta^2) \sinh \alpha b \sin \beta b = 0. \qquad (5.21)$$

Thus, for a given plate, we can obtain the natural frequency ω as contained in the expressions for α and β.

Example 5.3   Consider a plate whose edges $x = 0$ and $x = a$ are simply-supported, side $y = 0$ is also simply-supported, and side $y = b$ is free. Then, in terms of $W(x, y)$, the boundary conditions are

$$W = 0, \quad \frac{\partial^2 W}{\partial y^2} + \nu \frac{\partial^2 W}{\partial x^2} = 0 \qquad \text{(for } y = 0\text{)}, \qquad (5.22)$$

$$\frac{\partial^2 W}{\partial y^2} + \nu \frac{\partial^2 W}{\partial x^2} = 0, \quad \frac{\partial^3 W}{\partial y^3} + (2 - \nu) \frac{\partial^3 W}{\partial x^2 \partial y} = 0 \qquad \text{(for } y = b\text{)}. \qquad (5.23)$$

Equations (5.22) and (5.23) can now be expressed in terms of $Y_0(y)$ by substituting for $W(x, y)$ from (5.15). The boundary conditions (5.22) are satisfied if we take the general solution (5.17) which leads to

$$C_1 = C_3 = 0. \qquad (5.24)$$

Then the function $Y_0(y)$ may be written in the form

$$Y_0(y) = C_2 \sinh \alpha y + C_4 \sin \beta y. \qquad (5.25)$$

From the boundary conditions (5.23), it follows that

$$C_2(\alpha^2 + \nu \frac{m^2\pi^2}{a^2}) \sinh \alpha b - C_4(\beta^2 + \nu \frac{m^2\pi^2}{a^2}) \sin \beta b = 0,$$

$$C_2\alpha[\alpha^2 - (2-\nu)\frac{m^2\pi^2}{a^2}] \cosh \alpha b - C_4\beta[\beta^2 + (2-\nu)\frac{m^2\pi^2}{a^2}] \cos \beta b = 0.$$

(5.26)

For a nontrivial solution, the determinant of the coefficients of C_2 and C_4 should be zero. This results in the characteristic equation

$$\beta(\alpha^2 - \nu \frac{m^2\pi^2}{a^2}) \tanh \alpha b = \alpha(\beta^2 + \nu \frac{m^2\pi^2}{a^2}) \tan \beta b.$$

(5.27)

Since α and β contain the natural frequency ω, Eq. (5.27) can be used for the calculation of ω if the dimensions of the plate, the density, and the material properties are known. The first natural frequency is obtained by assuming $m = 1$ which is only one half wave along the X-direction.

5.2.2 FREE VIBRATION OF ORTHOTROPIC RECTANGULAR PLATES

In Section 5.2, we assumed that the material properties are the same in all the directions. However, in many practical applications, the material of the plates is anisotropic, e.g., wood, plywood, and fibre-reinforced plastics. In some cases, a structure becomes anisotropic by the presence of ribs or corrugations.

The orthotropic plates are a type of anisotropic plate. As stated in Section 3.5, orthotropic plates require four elastic constants E_x, E_y, ν_x, and G_{xy} for describing the orthotropic stress-strain relationships.

The governing differential equation for the buckling analysis of plates is given by Eq. (3.60). In a free vibration problem, the governing equation of motion is obtained by appending the inertia term due to lateral translation and setting equal to zero the terms due to inplane forces. Then we have

$$D_x \frac{\partial^4 w}{\partial x^4} + 2H \frac{\partial^4 w}{\partial x^2 \partial y^2} + D_y \frac{\partial^4 w}{\partial y^4} = -m_p \frac{\partial^2 w}{\partial t^2},$$

(5.28)

where $2H = D_x\nu_y + D_y\nu_x + 4D_{xy}$, and the moment-curvature relations are given by

$$M_x = -D_x(\frac{\partial^2 w}{\partial x^2} + \nu_y \frac{\partial^2 w}{\partial y^2}),$$

(5.29a)

$$M_y = -D_y(\frac{\partial^2 w}{\partial y^2} + \nu_x \frac{\partial^2 w}{\partial x^2}),$$

(5.29b)

$$M_{xy} = -2D_{xy}(\frac{\partial^2 w}{\partial x \partial y}).$$

(5.29c)

Example 5.4   Consider a plate which is simply-supported at all the four edges. The boundary conditions and the differential equation are satisfied by a solution of the form

$$w(x, y, t) = \sum_{m=1}^{\infty} \sum_{n=1}^{\infty} W_{mn} \sin \frac{m\pi x}{a} \sin \frac{n\pi y}{b} e^{i\omega t} \quad (m, n = 1, 2, 3, \ldots).$$

(5.30)

Substituting Eqs. (5.30) in Eq. (5.28), we obtain

$$D_x(\frac{m\pi}{a})^4 + 2H(\frac{mn\pi^2}{ab})^2 + D_y(\frac{n\pi}{b})^4 - \omega^2 m_p = 0.$$

(5.31)

Hence,

$$\omega_{mn} = \frac{\pi^2}{b^2}(\frac{1}{m_p})^{1/2} [D_x(\frac{mb}{a})^4 + 2Hn^2(\frac{mb}{a})^2 + D_y n^4]^{1/2}, \qquad (5.32)$$

and, by taking $m = n = 1$, we obtain the lowest frequency

$$\omega_{11} = \frac{\pi^2}{a^2}(\frac{1}{m_p})^{1/2} [D_x(\frac{b}{a})^4 + 2H(\frac{b}{a})^2 + D_y]^{1/2}. \qquad (5.33)$$

5.2.3   VIBRATION OF VISCOELASTIC PLATES

Let us consider another class of plates whose material behaves viscoelastically. As stated in Section 4.2.2, by making use of the correspondence principle, the equations governing the vibrations of a viscoelastic plate can be derived.

The governing equation for forced vibration of a perfectly elastic plate is

$$D\nabla^4 w + m_p \frac{\partial^2 w}{\partial t^2} = q(x, y, t) \qquad (5.34)$$

and the moment-curvature relations are

$$M_x = -D(\frac{\partial^2 w}{\partial x^2} + \nu \frac{\partial^2 w}{\partial y^2}),$$

$$M_y = -D(\frac{\partial^2 w}{\partial y^2} + \nu \frac{\partial^2 w}{\partial x^2}), \qquad (5.35)$$

$$M_{xy} = -D(1 - \nu) \frac{\partial^2 w}{\partial x \, \partial y}.$$

Taking the Laplace transform of Eqs. (5.34) and (5.35), we have

$$\bar{D}\nabla^4 \bar{w} + m_p[p^2\bar{w} - pw(x, y, 0) - \frac{dw}{dt}(x, y, 0)] = \bar{q}(x, y, p), \qquad (5.36)$$

where

$$\bar{w}(x, y, p) = \int_0^\infty e^{-pt} w(x, y, t) \, dt,$$

$$\bar{q}(x, y, p) = \int_0^\infty e^{-pt} q(x, y, t) \, dt, \qquad (5.37)$$

$$M_x = -\bar{D}(\frac{\partial^2 w}{\partial x^2} + \bar{\nu} \frac{\partial^2 w}{\partial y^2}),$$

$$M_y = -\bar{D}(\frac{\partial^2 w}{\partial y^2} + \bar{\nu} \frac{\partial^2 w}{\partial x^2}), \qquad (5.38)$$

$$M_{xy} = -\bar{D}(1 - \bar{\nu})(\frac{\partial^2 w}{\partial x \, \partial y}).$$

For perfectly elastic bodies, the relations that exist between Lamé's constants λ and μ and the material constants E, G, and ν are

$$G = \mu, \qquad \nu = \frac{\lambda}{2(\lambda + \mu)}, \qquad E = \frac{\mu(3\lambda + 2\mu)}{\lambda + \mu}. \qquad (5.39)$$

For viscoelastic bodies, D, ν are replaced by

$$\bar{D}(p) = 4\bar{\mu}(p)\frac{\bar{\lambda}(p) + \bar{\mu}(p)}{\bar{\lambda}(p) + 2\bar{\mu}(p)}\frac{h^3}{12},$$

$$\bar{\nu}(p) = \frac{\bar{\lambda}(p)}{2[\bar{\lambda}(p) + \bar{\mu}(p)]}.$$
(5.40)

For linear viscoelastic materials, the general stress-strain relations may be represented as

$$P_1(D)\sigma_{ij}(x, t) = P_2(D)\epsilon_{ij}(x, t),$$
(5.41a)

$$P_3(D)\sigma(x, t) = P_4(D)e(x,t),$$
(5.41b)

where

$$P_i(D) = \sum_{n=0}^{N_i} a_i^n D^n \quad (i = 1, 2, 3, 4)$$

are linear differential operators and $D^n(= \partial^n/\partial t^n)$ denotes the n-th time derivative. For convenience, Eqs. (5.41) may be rewritten in the form

$$P_1(D)P_3(D)\sigma_{ij} = P_2(D)P_3(D)\epsilon_{ij} + \tfrac{1}{3}\delta_{ij}[P_1(D)\cdot P_4(D) - P_2(D)\cdot P_3(D)]e.$$
(5.42)

In view of Eq. (5.42), we may express Eqs. (5.40) as

$$\bar{D}(p) = \frac{h^3}{12}\frac{P_2(p)}{P_1(p)}\frac{2P_1(p)P_4(p) + P_2(p)P_3(p)}{P_1(p)P_4(p) + 2P_2(p)P_3(p)},$$

$$\bar{\nu}(p) = \frac{P_1(p)P_4(p) - P_2(p)P_3(p)}{2P_1(p)P_4(p) + P_2(p)P_3(p)}.$$
(5.43)

The governing equation we get is

$$\frac{h^3}{12}P_2(D)[2P_1(D)P_4(D) + P_2(D)P_3(D)]\nabla^4 w + m_p P_1(D)[P_1(D)P_4(D) + 2P_2(D)P_3(D)]\frac{\partial^2 w}{\partial t^2}$$
$$= P_1(D)[P_1(D)P_4(D) + 2P_2(D)P_3(D)]q(x, y, t).$$
(5.44)

Now substituting Eqs. (5.43) in Eq. (5.36), we get the governing equation

$$\mu^2(p)\nabla^4\bar{w} + (p^2 w - pf + g) = \frac{\bar{q}}{m_p}, \qquad \mu^2(p) = \frac{\bar{D}(p)}{m_p},$$
(5.45)

with the initial conditions

$$w(x, y, 0) = f(x, y), \qquad \frac{\partial w}{\partial t}(x, y, 0) = g(x, y).$$
(5.46)

If we know the mathematical model for viscoelastic material, we can obtain the response in the same manner as described in Section 4.2.2.

5.3 ENERGY APPROACH

The method described in Section 5.2 is applicable in the study of plate problems, provided a closed-form solution can be obtained. This is possible only for certain sets of boundary conditions and for uniform plates. In practice, since a closed-form solution is not always attainable, we have to seek approximate solutions which are upper bounds to the exact solutions and can be derived by approaches based on certain energy principles. Although the governing equation for plates can also be derived from energy considerations, the energy principles are used here to obtain approximate analytical solutions.

The total potential energy π_p of a vibrating structural system is

$$\pi_p = U + V + T, \tag{5.47}$$

where U represents the strain energy of the plate, V the potential energy of the external forces, and T the kinetic energy.

For a vibrating plate, the kinetic energy T is given by

$$T = \tfrac{1}{2} \int_{\text{area}} m_p(x, y)(\frac{\partial w}{\partial t})^2 \, dx \, dy, \tag{5.48}$$

and the expressions for U and V are given by Eqs. (3.41) and (3.42), respectively.

5.3.1   RAYLEIGH'S METHOD

Rayleigh's method involves the determination of the kinetic and potential energies of a structural system by using assumed shape functions which satisfy the geometric (or kinematic) boundary conditions of the plate. For a conservative system, the maximum kinetic energy must equal the maximum potential energy. Thus, for free flexural vibrations of a plate, we have

$$U_{\text{b max}} = T_{\text{max}}, \tag{5.49}$$

where $U_{\text{b max}}$ is the maximum bending energy of the plate. Assuming the plate to be undergoing harmonic oscillations, we express the lateral deflection as

$$w(x, y, t) = W(x, y) \sin \omega t, \tag{5.50}$$

where $W(x, y)$ is the shape function and ω the natural frequency corresponding to the particular mode shape. Substituting relation (5.50) in Eqs. (5.48) and (3.41), we obtain

$$T_{\text{max}} = \frac{\omega^2}{2} \int_{\text{area}} m_p(x, y) W^2(x, y) \, dx \, dy = \frac{\omega^2}{2} I, \tag{5.51}$$

$$U_{\text{b max}} = \tfrac{1}{2} \int_{\text{area}} [D(\frac{\partial^2 W}{\partial x^2} + \frac{\partial^2 W}{\partial y^2})^2 + 2(1 - \nu)\{(\frac{\partial^2 W}{\partial x \, \partial y})^2 - \frac{\partial^2 W}{\partial x^2} \frac{\partial^2 W}{\partial y^2}\}] \, dx \, dy. \tag{5.52}$$

From Eq. (5.49), we have

$$\omega^2 = \frac{2 U_{\text{b max}}}{\int_{\text{area}} m_p(x, y) W^2(x, y) \, dx \, dy}. \tag{5.53}$$

Example 5.5   Consider a rectangular plate clamped on all edges, as shown in Fig. 5.2. The problem becomes doubly symmetric if we transfer the origin of the coordinate system to the centre of the plate. Then the boundary conditions of the plate are

$$W = 0, \quad \partial W/\partial x = 0 \quad (x = \pm a),$$
$$W = 0, \quad \partial W/\partial y = 0 \quad (y = \pm b). \tag{5.54}$$

The shape function which satisfies the geometric boundary conditions and closely approximates the first mode of vibration is

$$W(x, y) = C(x^2 - a^2)^2(y^2 - b^2)^2, \tag{5.55}$$

where C is a constant.

For a special case of rectangular plate, namely, a square plate, the substitution of

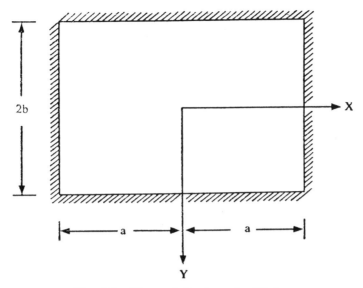

Fig. 5.2 Clamped Rectangular Plate.

Eq. (5.55) in Eqs. (5.51) and (5.52) and subsequent integration yield

$$\omega_1 = \frac{9.09}{a^2} \left(\frac{D}{m_p}\right)^{1/2}. \tag{5.56}$$

5.3.2 RAYLEIGH-RITZ METHOD

Ritz extended the Rayleigh method by including more than one parameter in the expression for the shape function. In so doing, not only was a more accurate value of the first natural frequency gained, but higher frequencies were also obtained. However, for accurate values of higher frequencies, more terms have to be included in the shape function.

Assuming the shape function $W(x, y)$ in the form of a series, we get

$$W(x, y) = A_1\phi_1(x, y) + A_2\phi_2(x, y) + \ldots + A_n\phi_n(x, y), \tag{5.57}$$

where $\phi_1(x, y)$, $\phi_2(x, y)$, $\ldots$, $\phi_n(x, y)$ are displacement functions which individually satisfy the geometric boundary conditions. If these are chosen as beam characteristic functions which are orthogonal, the convergence would be faster and thus affect the computational effort. The unknown coefficients $(A_1, A_2, \ldots, A_n)$ in Eq. (5.57) are obtained from the minimum total energy principle, i.e.,

$$\partial \pi_p / \partial A_i = 0 \qquad (i = 1, 2, \ldots, n), \tag{5.58}$$

where $\pi_p = U_{b\ max} - T_{max}$. From Eqs. (5.58), we obtain

$$\frac{\partial U_b}{\partial A_1} - \tfrac{1}{2}\omega^2 \frac{I}{A_1} = 0$$

$$\frac{\partial U_b}{\partial A_2} - \tfrac{1}{2}\omega^2 \frac{I}{A_2} = 0$$

$$\vdots$$

$$\frac{\partial U_b}{\partial A_n} - \tfrac{1}{2}\omega^2 \frac{I}{A_n} = 0$$

where I is as defined in (5.51). This leads to a set of homogeneous, linear algebraic equations in A_i. For a nontrivial solution, the determinant of A_i must be zero; this leads to a characteristic equation from which the natural frequencies can be obtained.

5.4 NUMERICAL TECHNIQUES

The use of differential equations for solving plate problems is limited to simple plate geometry and boundary conditions. When these conditions are complex, the application of differential equations becomes tedious and even impossible. In such situations, numerical techniques, such as finite difference and finite element, are increasingly used because, unlike differential equations, they can be efficiently programmed for computer application and thus automated to the desired accuracy.

5.4.1 FINITE DIFFERENCE TECHNIQUE

The numerical treatment of differential equations can produce approximate results which are applicable in most practical problems. In the finite difference method, the derivatives in the differential equation are replaced by difference quantities at some pivotal points. In Section 3.6.1, we explained this technique to show how the approximate results for plate buckling are obtained. The method described in this section is the same, except that now the inertia load term has to be added.

Let us assume that the plate undergoes harmonic oscillation about its equilibrium position. The deflection of the plate at any time t may then be expressed as

$$w(x, y, t) = W(x, y) \sin \omega t. \tag{5.59}$$

The substitution of Eq. (5.59) in Eq. (5.3) leads to

$$\nabla^4 W(x, y) - \omega^2 \frac{m_p}{D} W(x, y) = 0. \tag{5.60}$$

The plate is divided into a large number of square meshes. We replace the derivatives by central difference expressions, as described in Section 3.6.1.   Equation (5.60) then becomes

$$\frac{1}{h^4} [20W_{i, j} - 8(W_{i+1, j} + W_{i-1, j} + W_{i, j+1} + W_{i, j-1})$$

$$+ 2(W_{i+1, j+1} + W_{i-1, j+1} + W_{i+1, j-1} + W_{i-1, j-1})$$

$$+ W_{i+2, j} + W_{i-2, j} + W_{i, j+2} + W_{i, j-2}] = \omega^2 \frac{m_p}{D} W_{i, j}. \tag{5.61}$$

Example 5.6   Consider a clamped square plate with a finite difference mesh, as depicted in Fig. 5.3. At point (1, 2), we have

$$20W_{1, 1} - 8(W_{2, 1} + W_{0, 1} + W_{1, 2} + W_{1, 0}) + 2(W_{2, 2} + W_{0, 2} + W_{2, 0} + W_{0, 0})$$

$$+ W_{3, 1} + W_{-1, 1} + W_{1, 3} + W_{1, -1} = \lambda^4 W_{1, 1}; \tag{5.62a}$$

and at point (2, 1), we get

$$20W_{2, 1} - 8(W_{3, 1} + W_{1, 1} + W_{2, 2} + W_{2, 0}) + 2(W_{3, 2} + W_{1, 2} + W_{3, 0} + W_{1, 0})$$

$$+ W_{4, 1} + W_{0, 1} + W_{2, 3} + W_{2, -1} = \lambda^4 W_{2, 1}. \tag{5.62b}$$

Equations similar to Eqs. (5.62) are written at all the interior points (see Fig. 5.3). In all these equations, $\lambda^4 = \omega^2 m_p / D$. The boundary conditions of the plate in finite difference form are

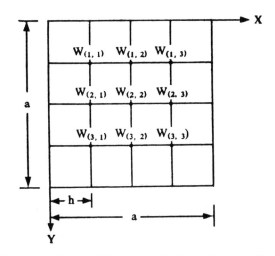

Fig. 5.3 Finite Difference Mesh for Square Plate.

$$W(0, j) = 0 \qquad (j = 1, 2, 3, 4),$$
$$W(4, j) = 0 \qquad (j = 1, 2, 3, 4),$$
$$W(i, 0) = 0 \qquad (i = 1, 2, 3, 4),$$
$$W(i, 4) = 0 \qquad (i = 1, 2, 3, 4),$$
$$W(-1, j) = W(1, j) \qquad (j = 1, 4),$$
$$W(i, -1) = W(i, +1) \qquad (i = 1, 4),$$
$$W(3, j) = W(5, j) \qquad (j = 1, 4),$$
$$W(i, 3) = W(i, 5) \qquad (i = 1, 4).$$

$$(5.63)$$

Using conditions (5.63), we can obtain the solution for the plate. However, this method does not converge fast. Hence, we shall now describe a modified finite difference technique.

5.4.2 IMPROVED FINITE DIFFERENCE TECHNIQUE

When higher-order derivatives of W and numerous mesh points are involved in the analysis of plates, the solution may converge to wrong values because of machine errors. The approximation of boundary conditions also causes errors. Besides, the large number of equations to be solved (because of the extremely fine mesh) may create round-off errors in the computer solutions.

The finite difference method can be improved upon by
    (i) replacing the derivatives with higher-order polynomials;
    (ii) refining the representation of the boundary conditions;
    (iii) amending the load averaging technique.

This is achieved by incorporating additional mesh points near a pivotal point i. Let us consider the derivation for an improved expression for the second derivative. To do this expand the function $f(x)$ at the points $(i + 1)$ and $(i - 1)$ as

$$f_{i+1} = f(x_i + h) = f(x_i) + \frac{h}{1!}\frac{df}{dx}\bigg|_i + \frac{(h)^2}{2!}\frac{d^2f}{dx^2}\bigg|_i + \cdots + \frac{(h)^k}{k!}\frac{d^kf}{dx^k}\bigg|_i, \qquad (5.64)$$

$$f_{i-1} = f(x_i - h) = f(x_i) - \frac{h}{1!}\frac{df}{dx}\bigg|_i + \frac{(h)^2}{2!}\frac{d^2f}{dx^2}\bigg|_i + \cdots + \frac{(-h)^k}{k!}\frac{d^kf}{dy^k}\bigg|_i + \cdots . \qquad (5.65)$$

By adding Eqs. (5.64) and (5.65), we get

$$\frac{d^2f}{dx^2}\bigg|_i = \frac{1}{(h)^2}(f_{i+1} - 2f_i + f_{i-1}) - \frac{(h)^2}{12}\frac{d^4f}{dy^4}\bigg|_i - \frac{(h)^4}{360}\frac{d^4f}{dy^4}\bigg|_i + \cdots . \tag{5.66}$$

By expressing the fourth-order derivative in terms of central differences, we obtain

$$\frac{d^2f}{dx^2}\bigg|_i = \frac{1}{12(h)^2}(-f_{i+2} + 16f_{i+1} - 30f_i + 16f_{i-1} - f_{i-2}) + \frac{(h)^4}{90}\frac{d^4f}{df^4}\bigg|_i . \tag{5.67}$$

The magnitude of the error in this case is of the order $O(h^4)$ which may be compared with $O(h^2)$ in the central difference formula used earlier. Similar expressions can be derived for higher-order derivatives. These can then be used in obtaining the finite difference solution.

5.4.3   FINITE ELEMENT TECHNIQUE

Often a vibrating plate has to be replaced by an equivalent discrete system. At such times, we can either employ the lumped mass system or the consistent mass system. The advantages and disadvantages of these techniques have been discussed in Section 4.5.2. A large number of elements are available for representing plate behaviour. We can, for instance, use triangular or rectangular plate elements with inplane or bending displacements. The elements are chosen on the basis of the nature of displacements, the geometry of the structure, and also the type of plate oscillation.

For the free vibration of plates, the governing equation is given by Eq. (4.101) with the load term as zero. In view of the size of the matrix employed for plate problems, computer techniques are resorted to for their solution.

The method for obtaining the stiffness matrix for plates has been discussed in Section 3.6.2. We shall now consider some examples for deriving the inertia matrix for various types of plate elements.

Rectangular Plate with Translational Displacement (Consistent mass matrix)

Consider a rectangular plate element with nodal degrees of freedom as shown in Fig. 5.4. Here,

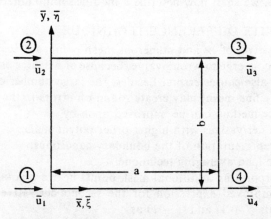

Fig. 5.4   Inplane Nodal Degrees of Freedom.

the origin of the local coordinate system is taken at the lower left corner of the rectangle. Defining the nondimensional parameters ξ and η as

$$\xi = x/a, \qquad \eta = y/b, \tag{5.68}$$

we get the displacement u_x at any point within the element in the form

$$\bar{u}_x = \alpha_1 \xi + \alpha_2 \xi \eta + \alpha_3 \eta + \alpha_4 \tag{5.69}$$

which satisfies the assumption of linearly-varying boundary displacements. The generalized coordinates α_1, α_2, α_3, and α_4 are written in terms of boundary displacements. The boundary conditions for the element are

$$\bar{u}_x(0, 0) = \bar{u}_1, \qquad \bar{u}_x(0, 1) = \bar{u}_2, \qquad \bar{u}_x(1, 1) = \bar{u}_3, \qquad \bar{u}_x(1, 0) = \bar{u}_4. \tag{5.70}$$

Substituting these boundary values in the equations for displacements, we have

$$\bar{u}_x = (1 - \xi)(1 - \eta)\bar{u}_1 + (1 - \xi)\eta\bar{u}_2 + \xi\eta\bar{u}_3 + \xi(1 - \eta)\bar{u}_4. \tag{5.71}$$

The matrix $[N]$ can therefore be written as

$$[N] = [(1 - \xi)(1 - \eta) \mid (1 - \xi)\eta \mid \xi\eta \mid \xi(1 - \eta)] \tag{5.72}$$

and the mass matrix $[m]$ is given by

$$[m] = \int_V \rho_p [N]^T [N] \, dv, \tag{5.73}$$

$$[m] = V\rho_p \int_0^1 \int_0^1 [N]^T [N] \, d\xi \, d\eta, \tag{5.74}$$

where V is the volume of the element. For a rectangular element, the consistent mass matrix $[m]$ may be written in matrix notation as

$$[m] = \frac{\rho_p V}{36} \begin{bmatrix} 4 & 2 & 1 & 2 \\ 2 & 4 & 2 & 1 \\ 1 & 2 & 4 & 2 \\ 2 & 1 & 2 & 4 \end{bmatrix}. \tag{5.75}$$

Rectangular Plate with Bending Displacements

In the bending of rectangular plates, two different displacement functions for the transverse displacement $\bar{u}_z$ are possible. The first is such that the boundary displacements on adjacent plate elements are compatible; however, since rotations of the element edges on a common boundary are not compatible, the slope is discontinuous [see Eq. (3.76)]. In the second, both deflection and slope continuity on adjacent elements are ensured. Here again, we can neglect

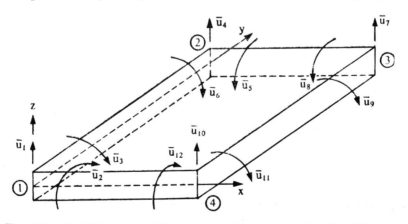

Fig. 5.5   Nodal Degrees of Freedom for Rectangular Bending Element.

the inertia forces due to the rotational degree of freedom and need to consider only the translational mass matrix.

Figure 5.5 shows a rectangular plate element with nodal degrees of freedom. For the first type of displacement, N^T is given by Eq. (3.76) and for the second type N^T is given by

$$
N^T = \begin{bmatrix}
(1 + 2\xi)(1 - \xi)^2(1 + 2\eta)(1 - \eta)^2 \\
(1 + 2\xi)(1 - \xi)^2\eta(1 - \eta^2)b \\
-\xi(1 - \xi)^2(1 + 2\eta)(1 - \eta)^2a \\
(1 + 2\xi)(1 - \xi)^2(3 - 2\eta)\eta^2 \\
-(1 + 2\xi)(1 - \xi)^2(1 - \eta)\eta^2 b \\
-\xi(1 - \xi)^2(3 - 2\eta)\eta^2 a \\
(3 - 2\xi)\xi^2(3 - 2\eta)\eta^2 \\
-(3 - 2\xi)\xi^2(1 - \eta)\eta^2 b \\
(1 - \xi)\xi^2(3 - 2\eta)\eta^2 a \\
(3 - 2\xi)\xi^2(1 + 2\eta)(1 - \eta)^2 \\
(3 - 2\xi)\xi^2\eta(1 - \eta)^2 b \\
(1 - \xi)\xi^2(1 + 2\eta)(1 - \eta)^2 a
\end{bmatrix}
\tag{5.76}
$$

If we know N^T, it is possible to obtain the mass matrix $[m]$.

Triangular Plate Element (Inplane forces)

Consider a constant strain triangular element (see Fig. 5.6), where the displacement variation

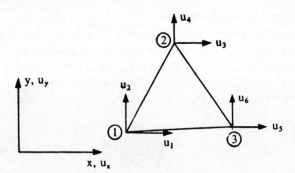

Fig. 5.6   Nodal Degrees of Freedom for Constant Strain Triangular Element.

within the element is given by

$$
\begin{aligned}
u_x &= \alpha_1 x + \alpha_2 y + \alpha_3, \\
u_y &= \alpha_4 x + \alpha_5 y + \alpha_6.
\end{aligned}
\tag{5.77}
$$

The generalized coordinates $\alpha_1, \alpha_2, \ldots, \alpha_6$ are determined from the displacements of the three vertices of the triangle. The displacements u_x and u_y may be written as

$$
u_x = \frac{1}{2A}[\{(y_3 - y_2)(x - x_2) - (x_3 - x_2)(y - y_2)\}u_1 + \{-(y_3 - y_1)(x - x_3)
$$
$$
+ (x_3 - x_1)(y - y_3)\}u_3 + \{(y_2 - y_1)(x - x_1) - (x_2 - x_1)(y - y_1)\}u_5],
\tag{5.78}
$$

$$u_y = \frac{1}{2A}[\{(y_3 - y_2)(x - x_2) - (x_3 - x_2)(y - y_2)\}u_2 + \{-(y_3 - y_1)(x - x_3)$$
$$+ (x_3 - x_1)(y - y_3)\}u_4 + \{(y_2 - y_1)(x - x_1) - (x_2 - x_1)(y - y_1)\}u_6], \qquad (5.79)$$

where A is the area of the triangle. From Eqs. (5.78) and (5.79), it is obvious that the displacements along any edge vary linearly and depend on only the displacements of the two vertices along that edge; this means that the compatibility of displacement along the adjacent edges is satisfied.

To find the matrix $[B]$, (5.78) and (5.79) may be incorporated in Eq. (2.89); this gives

$$\begin{Bmatrix} e_{xx} \\ e_{yy} \\ e_{yx} \end{Bmatrix} = \frac{1}{2A} \begin{bmatrix} (y_3-y_2) & 0 & -(y_3-y_1) & 0 & (y_2-y_1) & 0 \\ 0 & -(x_3-x_2) & 0 & (x_3-x_1) & 0 & -(x_2-x_1) \\ -(x_3-x_2) & (y_3-y_2) & (x_3-x_1) & -(y_3-y_1) & -(x_2-x_1) & (y_2-y_1) \end{bmatrix} \begin{Bmatrix} u_1 \\ \vdots \\ \vdots \\ u_6 \end{Bmatrix}. \qquad (5.80)$$

For plane stress problem, the numerical matrix $[\chi]$ is given by

$$[\chi] = \frac{E}{(1 - v^2)} \begin{bmatrix} 1 & v & 0 \\ v & 1 & 0 \\ 0 & 0 & \dfrac{1-v}{2} \end{bmatrix}. \qquad (5.81)$$

The substitution of expressions for $[B]$ and $[\chi]$ in Eq. (2.98) and subsequent integration lead to the stiffness matrix $[k]$. This can be separated into two parts, i.e.,

$$[k] = [k_n] + [k_s], \qquad (5.82)$$

where $[k_n]$ represents the stiffness due to normal stresses and k_s the stiffness due to shear stresses which are symmetric along the main diagonal. In matrix notation, these may be written as

$$[k_n] = \frac{Eh}{4A(1-v^2)} \times$$

$$\begin{bmatrix} (y_3-y_2)^2 \\ -v(y_3-y_2)(x_3-x_2) & (x_3-x_2)^2 \\ -(y_3-y_2)(y_3-y_1) & v(x_3-x_2)(x_3-x_1) & (y_3-y_1)^2 \\ v(y_3-y_2)(x_3-x_1) & -(x_3-x_2)(x_3-x_1) & -v(y_3-y_1)(x_3-x_1) & (x_3-x_1)^2 \\ (y_3-y_2)(y_2-y_1) & -v(x_3-x_2)(y_2-y_1) & -(y_3-y_1)(y_2-y_1) & v(x_3-x_1)(y_2-y_1) & (x_2-x_1)^2 \\ -v(y_3-y_2)(x_2-x_1) & (x_3-x_2)(x_2-x_1) & v(y_3-y_1)(x_2-x_1) & -(x_3-x_1)(x_2-x_1) & -v(y_2-y_1)(x_2-x_1)^2 & (x_2-x_1)^2 \end{bmatrix},$$
$$(5.83)$$

$$[k_s] = \frac{Eh}{8A(1+v)} \times$$

$$\begin{bmatrix} (x_3-x_2)^2 \\ -(x_3-x_2)(y_3-y_2) & (y_3-y_2)^2 \\ -(x_3-x_2)(x_3-x_1) & (y_3-y_2)(x_3-x_1) & (x_3-x_1)^2 \\ (x_3-x_2)(y_3-y_1) & -(y_3-y_2)(y_3-y_1) & -(x_3-x_1)(y_3-y_1) & (y_3-y_1)^2 \\ (x_3-x_2)(x_2-x_1) & -(y_3-y_2)(x_2-x_1) & -(x_3-x_1)(x_2-x_1) & (y_3-y_1)(x_2-x_1) & (x_2-x_1)^2 \\ -(x_3-x_2)(y_2-y_1) & (y_3-y_2)(y_2-y_1) & (x_3-x_1)(y_2-y_1) & -(y_3-y_1)(y_2-y_1) & -(x_2-x_1)(y_2-y_1) & (y_2-y_1)^2 \end{bmatrix}.$$
$$(5.84)$$

For a lumped mass system, the area for each node is taken as one-third the element area. Hence, the lumped mass matrix may be represented as

$$[m] = \frac{\rho_{\mathrm{p}} h A}{3} \begin{bmatrix} 1 & 0 & 0 & 0 & 0 & 0 \\ 0 & 1 & 0 & 0 & 0 & 0 \\ 0 & 0 & 1 & 0 & 0 & 0 \\ 0 & 0 & 0 & 1 & 0 & 0 \\ 0 & 0 & 0 & 0 & 1 & 0 \\ 0 & 0 & 0 & 0 & 0 & 1 \end{bmatrix}. \tag{5.85}$$

If we know $[k]$ and $[m]$ for the element, we can obtain the global stiffness and global mass matrices. Then we can get the natural frequencies and the associated mode shapes.

BIBLIOGRAPHY

Bathe, K. J. and Wilson, E., Numerical Methods in Finite Element Analysis, Prentice-Hall, Englewood Cliffs, New Jersey, 1976.

Desai, C. S. and Abel, J. F., Introduction to the Finite Element Method, Van Nostrand Reinhold, New York, 1972.

Marguerre, K. and Woernle, H. T., Elastic Plates, Blaisdell, Waltham, Mass., 1969.

Martin, H. C. and Carey, G. F., Introduction to Finite Element Analysis, McGraw-Hill, New York, 1973.

Nowacki, W., Dynamics of Elastic Systems, Chapman and Hall, London, 1963.

Salvadori, M. G. and Baron, M. L., Numerical Methods in Engineering, Prentice-Hall, Englewood Cliffs, New Jersey, 1961.

Zienkiewicz, O. C., The Finite Element Method in Engineering Science, 2nd edn., McGraw-Hill, London, 1971.

6

Random Vibration

6.1 INTRODUCTION

The theory of random vibration provides a probabilistic treatment of uncertainty in vibration problems. Most physical systems operate in random vibration environment, e.g., flight vehicles subjected to gust loading, jet engine noise, and boundary layer turbulence; aircraft; motor vehicles; and trains subjected to ground-induced excitation. In this chapter, an elementary analysis of the theory of random vibration is given after a brief review of the theory of random processes.

6.2 RANDOM PROCESSES

In probability theory (it is assumed that the reader is familiar with the fundamentals), a random variable is defined as the uncertain outcome of a conceptual experiment. A set containing all possible outcomes of such an experiment is called the *sample space*. Let the outcome of a conceptual or real experiment be the function of a parameter, say time t. Each realization of the experiment is then a function of time, known as a *sample function*, and repeated realizations of

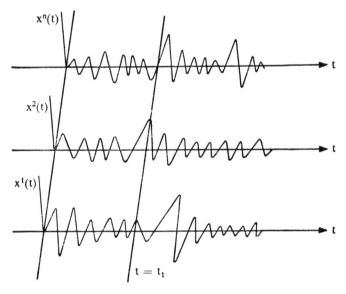

Fig. 6.1 Ensemble of Random Process.

the same experiment consist of a family of functions called the samples of a *random process* $X(t)$, as shown in Fig. 6.1. We say that $x^1(t)$, $x^2(t)$, ..., $x^n(t)$ are sample functions of the random process $X(t)$, where t is defined over an interval $t \in T : (0, a]$, and is called the indexing parameter over the indexing set T. It should be noted that, at a fixed time t_1, $X(t_1)$ is a random variable and $x^1(t_1)$, $x^2(t_1)$, ... are its realizations. Examples of random processes are

      (i) pressure fluctuations occurring with time at a point on a missile due to jet engine noise;

      (ii) ground acceleration at a point during an earthquake.

A random process may be characterized by one or more indexing parameters. It is then called a multiparametered random process, for instance, the load on a beam (see Fig. 6.2) varying at random with the distance along the beam and time, i.e.,

$$X(t, z), \qquad t \in T : (0, a], z \in Z : (0, L],$$

is a two-parameter random process.

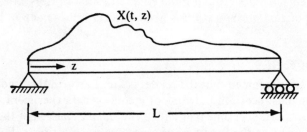

Fig. 6.2 Beam with Random Loading.

In view of the foregoing discussion, a random process may be defined as a parametered family of random variables with the parameter (or parameters) belonging to an indexing set (or sets).

In this chapter, we shall confine our discussion to single parameter families, usually a function of time, and call them *random excitations*. Noting that a random variable may be discrete or continuous, and that a random process parameter may also be discrete or continuous, we can classify random processes into

      (i) continuously parametered continuous random process;

      (ii) continuously parametered discrete random process;

      (iii) discretely parametered continuous random process;

      (iv) discretely parametered discrete random process.

Probability Description of Random Processes

A random process is characterized by a hierarchy of joint probability density functions of the form

$$p_{\{x\}}(x_1, t_1)$$
$$p_{\{x\}}(x_1, t_1; x_2, t_2)$$
$$\vdots$$
$$p_{\{x\}}(x_1, t_1; x_2, t_2; \ldots; x_n, t_n)$$
$$\vdots$$

(6.1)

where $x_i = x(t_i)$. A random process can also be specified by a hierarchy of characteristic functions

$$M_{\{X\}}(\theta_1, t_1) = E[\exp\{i\theta_1 x(t_1)\}]$$
$$M_{\{X\}}(\theta_1, t_1; \theta_2, t_2) = E[\exp\{i\theta_1 x(t_1) + i\theta_2 x(t_2)\}]. \qquad (6.2)$$
$$\vdots$$

Moments   Moment functions for a random process are defined in the same manner as for a random variable.

Mean

$$\mu_X(t) = E[X(t)] = \int_{-\infty}^{\infty} x p_{\{X\}}(x, t)\, dx. \qquad (6.3)$$

Autocorrelation function

$$\phi_{XX}(t_1, t_2) = E[X(t_1)X(t_2)] = \int_{-\infty}^{\infty}\int_{-\infty}^{\infty} x_1 x_2 p_{\{X\}}(x_1, t_1; x_2, t_2)\, dx_1\, dx_2. \qquad (6.4)$$

Autocovariance

$$K_{XX}(t_1, t_2) = E[\{(X(t_1) - \mu_X(t_1)\}\{X(t_2) - \mu_X(t_2)\}] = \phi_{XX}(t_1, t_2) - \mu_X(t_1)\mu_X(t_2). \qquad (6.5)$$

It should be noted that

$$K_{XX}(t, t) = E[\{X(t) - \mu_X(t)\}^2] = \sigma_X^2(t).$$

Crosscovariance

$$K_{XY}(t_1, t_2) = E[\{X(t_1) - \mu_X(t_1)\}\{Y(t_2) - \mu_Y(t_2)\}] = \phi_{XY}(t_1, t_2) - \mu_X(t_1)\mu_Y(t_2). \qquad (6.6)$$

Normalized covariance

$$\rho_{XX}(t_1, t_2) = K_{XX}(t_1, t_2)/[\sigma_X(t_1)\sigma_X(t_2)]. \qquad (6.7)$$

In the foregoing moment functions, the prefix "auto" refers to the same random process, and the prefix "cross" indicates different random processes. From the definition of correlation and covariance functions, it is clear that these functions are symmetric with respect to t_1 and t_2, i.e.,

$$\phi_{XX}(t_1, t_2) = \phi_{XX}(t_2, t_1),$$
$$K_{XX}(t_1, t_2) = K_{XX}(t_2, t_1), \qquad (6.8)$$
$$K_{XY}(t_1, t_2) = K_{YX}(t_2, t_1) \neq K_{YX}(t_1, t_2).$$

Moment functions describe the gross properties of a given random phenomenon. In theory, all the moment functions are needed to fully describe the probability structure of a given random process. In practice, however, only the first few, namely, the mean and correlation functions are used.

Stationary Random Process
A random process is said to be stationary if its probability structure is independent of the shift of parametric origin. In Eqs. (6.1), shifting the origin by $-t_1$, for a stationary random process, we get

$$p_{\{X\}}(x_1, t_1) = p_{\{X\}}(x_1, 0) \qquad \text{[independent of time]} \qquad (6.9)$$
$$p_{\{X\}}(x_1, t_1; x_2, t_2) = p_{\{X\}}(x_1, 0; x_2, t_2 - t_1) \qquad \text{[function of difference } (t_2 - t_1)] \qquad (6.10)$$
$$\vdots$$
$$p_{\{X\}}(x_1, t_1; \ldots; x_n, t_n) = p_{\{X\}}(x_1, 0; \ldots; x_n, t_n - t_1) \qquad (6.11)$$
$$\vdots$$

If this property holds for only the first and second probability densities, the random process is called weakly stationary, i.e., stationary in a general sense.

From the definition of moment functions, i.e., Eqs. (6.3)–(6.6) and properties (6.9) and (6.10), it is clear that, for a stationary random process,

$$\mu_X(t) = \mu_X = \text{constant,} \tag{6.12}$$

$$\phi_{XX}(t_1, t_2) = R_{XX}(t_1 - t_2) = R_{XX}(\tau), \tag{6.13}$$

$$K_{XX}(t_1, t_2) = \Gamma_{XX}(t_1 - t_2) = \Gamma_{XX}(\tau), \tag{6.14}$$

$$K_{XY}(t_1, t_2) = \Gamma_{XY}(t_1 - t_2) = \Gamma_{XY}(\tau), \tag{6.15}$$

where $R_{XX}(\tau)$ is the autocorrelation function, $\Gamma_{XX}(\tau)$ the autocovariance function, and $\Gamma_{XY}(\tau)$ the crosscovariance function.

Properties of Correlation Function

(i) From the symmetry property given by (6.5), it is obvious that

$$R_{XX}(t_1 - t_2) = R_{XX}(t_2 - t_1), \qquad R_{XX}(\tau) = R_{XX}(-\tau). \tag{6.16}$$

Hence, $R_{XX}(\tau)$ is an even function of τ.

(ii) $R_{XX}(\tau)$ is a positive definite function.

(iii)

$$|R_{XY}(\tau)| \leqslant \sqrt{R_{XX}(0)R_{YY}(0)}. \tag{6.17}$$

Therefore, $|R_{XX}(\tau)| \leqslant R_{XX}(0)$ so that $R_{XX}(\tau)$ is maximum at $\tau = 0$; thus,

$$\frac{d}{d\tau}R_{XX}(\tau)\bigg|_{\tau=0} = 0, \qquad \frac{d^2}{d\tau^2}R_{XX}(\tau)\bigg|_{\tau=0} \leqslant 0. \tag{6.18}$$

(iv) If a random process $X(t)$ does not contain periodic components, then

$$R_{XX}(\tau) = 0 \qquad (\tau \to \infty). \tag{6.19}$$

A typical autocorrelation function is shown in Fig. 6.3.

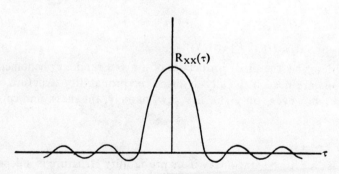

Fig. 6.3   Typical Autocorrelation Function.

Temporal Averages

For a given sample function $x^j(t)$, we can define averages over time, called *temporal averages,* as

$$\langle X(t) \rangle = \lim_{T \to \infty} \frac{1}{T} \int_0^T x(t)\, dt, \tag{6.20}$$

$$\langle X^2(t) \rangle = \lim_{T \to \infty} \frac{1}{T} \int_0^T x^2(t) \, dt, \tag{6.21}$$

$$\langle X(t)X(t+\tau) \rangle = \lim_{T \to \infty} \frac{1}{T-\tau} \int_0^{T-\tau} x(t+\tau)x(t) \, dt. \tag{6.22}$$

A subclass of stationary processes is said to be *ergodic in the mean* if

$$\langle X \rangle = E[X] = \mu_X, \tag{6.23}$$

and *ergodic in correlation* if

$$\langle X(t)X(t+\tau) \rangle = E[X(t)X(t+\tau)] = R_{XX}(\tau). \tag{6.24}$$

We may recall that the expected values $E[f(X)]$ are the ensemble averages of a random process. Hence, the ergodic property holds if ensemble averages and temporal averages are equal.

It may be noted that the ergodic property permits us to get moment functions of a stationary random process from a single long record which contains all the statistical information about the random phenomenon. When there are only one or two sample functions, ergodicity is commonly assumed. When more information is available, the ergodicity assumption can be verified, for instance, in the designing of flight vehicles.

Operations on Random Processes

Mathematical models of physical systems subjected to random excitation involve derivatives and integrals of random processes. These operations are defined in terms of probability, since the variables characterizing random processes are probabilistic. To do this, we need to define first the concepts of limit and continuity. These concepts are linked with convergence which is defined in several ways. Here, we shall confine our discussion to the notion of *mean square convergence*.

Limit   A random sequence X_n is said to converge in the mean square to a random variable X if

$$\lim_{n \to \infty} E[|X_n - X|^2] = 0. \tag{6.25}$$

This may be abbreviated to

$$\text{l.i.m.}_{n \to \infty} X_n = X.$$

If Eq. (6.25) holds, it is necessary that $E[X_n^2] < \infty$. A random process $X(t)$ satisfying $E[X^2(t)] < \infty$ converges in the mean square if and only if $\phi_{XX}(t, t')$ exists and is finite regardless of the convergence of t, t'. Thus, it can be shown that

$$\lim_{t \to t_0} E[X(t)] = E[\text{l.i.m.}_{t \to t_0} X(t)] = E[X]. \tag{6.26}$$

Continuity   A random process $X(t)$ is said to be *continuous in the mean* if

$$\text{l.i.m.}_{h \to 0} X(t+h) = X(t). \tag{6.27}$$

A necessary and sufficient condition for Eq. (6.27) is

$$\phi_{XX}(t, s) < \infty \text{ and continuous on } t = s. \tag{6.28}$$

Derivative   A random process $X(t)$ is described as *differentiable in the mean* if

$$\frac{d}{dt}X(t) = \dot{X}(t) = \text{l.i.m.}_{h \to 0} \frac{X(t+h) - X(t)}{h} \tag{6.29}$$

exists. A necessary and sufficient condition for Eq. (6.29) is

$$\frac{\partial^2 \phi_{XX}(t, s)}{\partial t \, \partial s} < \infty \quad \text{and continuous on } t = s. \tag{6.30}$$

It can be shown that if Eq. (6.30) holds, then

$$\frac{d}{dt} E[X(t)] = E[\frac{d}{dt} X(t)] = E[\dot{X}(t)], \tag{6.31}$$

that is, expectation and derivative operations commute. Similarly, it can be shown that if $X(t)$ is weakly stationary, then

$$R_{\dot{X}X}(t - s) = \dot{R}_{XX}(t - s), \qquad R_{\dot{X}\dot{X}}(t - s) = -\ddot{R}_{XX}(t - s). \tag{6.32}$$

Since $\dot{R}_{XX}(\tau) = 0$, Eqs. (6.18) become

$$R_{\dot{X}X}(\tau) = 0. \tag{6.33}$$

Thus, $\dot{X}$ and X are orthogonal.

Integration   Let $X(t)$ be a random process and $h(t, \tau)$ a bounded deterministic function. Then $Z(\tau)$, defined by

$$Z(\tau) = \int_a^b X(t) h(t, \tau) \, dt, \tag{6.34}$$

is called a *transformed random process* which exists in the mean as a limit of the Riemann sum if and only if

$$\left| \int_a^b \int_a^b \phi_{XX}(t_1, t_2) h(t_1, \tau_1) h(t_2, \tau_2) \, dt_1 \, dt_2 \right| < \infty \quad \text{for all } \tau_1 \text{ and } \tau_2. \tag{6.35}$$

If inequality (6.35) is true, then it can be shown that

$$E[Z(\tau)] = \int_a^b h(t, \tau) E[X(t)] \, dt, \tag{6.36}$$

$$E[Z(\tau_1)Z(\tau_2)] = \int_a^b \int_a^b h(t_1, \tau_1) h^*(t_2, \tau_2) E[X(t_1)X(t_2)] \, dt_1 \, dt_2, \tag{6.37}$$

where $h^*(t, \tau)$ is the complex conjugate of $h(t, \tau)$. In other words, expectation and integration operations commute.

Spectral Decomposition of Random Processes

The Fourier transform of the autocorrelation function of a stationary random process $X(t)$ is called the *power spectral density* (psd) and is given by

$$\Phi_{XX}(\omega) = \frac{1}{2\pi} \int_{-\infty}^{\infty} R_{XX}(\tau) e^{-i\omega\tau} \, d\tau. \tag{6.38}$$

By the inversion theorem, we get

$$R_{XX}(\tau) = \int_{-\infty}^{\infty} \Phi_{XX}(\omega) e^{i\omega\tau} \, d\tau, \tag{6.39}$$

$$R_{XX}(0) = E[X^2(t)] = \int_{-\infty}^{\infty} \Phi_{XX}(\omega) \, d\omega. \tag{6.40}$$

If $X(t)$ is identified with the displacement of a physical system, $R_{XX}(0)$ represents the average energy. From Eq. (6.40), it is clear that $\Phi_{XX}(\omega)$ stands for the energy density associated with the frequency ω. Thus, $\Phi_{XX}(\omega)$ represents the spectral distribution of energy in a system. This information is useful in structural design, as described in Chapter 17. Since $R_{XX}(\tau)$ is an even function of τ, it can be shown that $\Phi_{XX}(\omega)$ is a real even function of ω and

$$\Phi_{XX}(\omega) = \frac{1}{\pi} \int_0^\infty R_{XX}(\tau) \cos \omega\tau \, d\omega, \qquad (6.41)$$

$$R_{XX}(\tau) = 2 \int_0^\infty \Phi_{XX}(\omega) \cos \omega\tau \, d\omega, \qquad (6.42)$$

which are called Wiener-Khintchine relations.

Sometimes it is convenient to work with frequencies in cycles/sec. With this definition, we may rewrite (6.40) as

$$R_{XX}(0) = E[X^2(t)] = 4\pi \int_0^\infty \Phi_{XX}(\omega) \, dp = \int_0^\infty \Phi_{XX}^*(p) \, dp, \qquad (6.43)$$

where $\Phi_{XX}^*(p) = 4\pi\Phi_{XX}(\omega)$, ω being equal to $2\pi p$.

Classification of Random Processes

Taking into account the fact that the energy of stationary random processes is distributed over the frequency ω, we can classify the random processes into two categories.

Broad band random processes   A random process is said to be of the broad band random type if $\Phi_{XX}(\omega)$ has significant values over a wide range of frequencies.

A typical sample function and the corresponding $\Phi_{XX}(\omega)$ and $R_{XX}(\tau)$ for a broad band process are shown in Fig. 6.4. Examples of the broad band random process are

(i) pressure fluctuations on the surface of a flight vehicle due to boundary layer turbulence and jet engine noise;

(ii) gust loading;

(iii) ground acceleration during earthquakes.

A common idealization of the broad band random process is to assume $\Phi_{XX}(\omega) = \Phi_0 =$ constant (see Fig. 6.5) which gives

$$R_{XX}(\tau) = 2\pi\Phi_0\delta(\tau). \qquad (6.44)$$

Such a random process is called *white noise*. We should note that since for white noise $R_{XX}(0) \to \infty$, physically it is not a realizable idealization. Our subsequent discussion shows that, under suitable conditions, white noise gives a meaningful analytical response of systems.

Another idealization of the broad band process is a *band-limited white noise* given by

$$\Phi_{XX}(\omega) = \Phi_0 \qquad (\omega_1 < |\omega| < \omega_2)$$
$$= 0 \qquad \text{(otherwise)}, \qquad (6.45)$$

$$R_{XX}(\tau) = 2\Phi_0 \frac{\sin \omega_2\tau - \sin \omega_1\tau}{\tau}. \qquad (6.46)$$

For example, electrodynamic random shakers can be set to give a spectrum, as shown in Fig. 6.6.

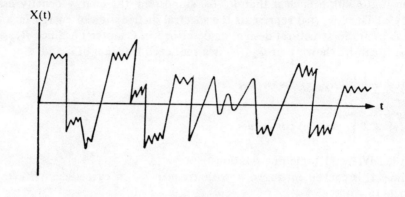

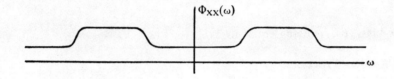

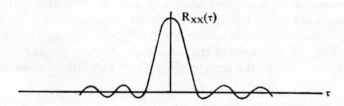

Fig. 6.4   Broad Band Random Process.

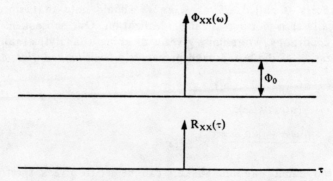

Fig. 6.5   Schematic Representation of White Noise.

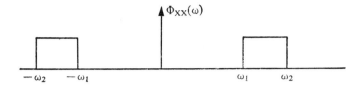

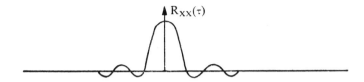

Fig. 6.6   Band-Limited White Noise.

Narrow band random processes   A random process is said to be of the narrow band type if $\Phi_{XX}(\omega)$ has significant values over a narrow band around a central frequency.

A typical sample function and its corresponding $\Phi_{XX}(\omega)$ and $R_{XX}(\tau)$ for a narrow band process are shown in Fig. 6.7. Examples of the narrow band random process are

(i) response of a strongly resonant linear system to the broad band random process;

(ii) dynamic unbalance.

Chebyshev Inequality

For a random variable X, the Chebyshev inequality gives

$$P(|X - \mu_X| \geqslant h\sigma_X) \leqslant 1/h^2. \tag{6.47}$$

If $X(t)$, $t \in T : [a, b]$, is a random process, it can be shown that

$$P(|X(t) - \mu_X(t)| \geqslant \epsilon \text{ for some } t \in T : [a, b])$$

$$\leqslant \frac{1}{2\epsilon} [\sigma_X^2(a) + \sigma_X^2(b)] + \frac{1}{\epsilon^2} \int_a^b \sigma_X(t)\sigma_{\dot{X}}(t) \, dt \qquad (\epsilon > 0). \tag{6.48}$$

This is called the generalized Chebyshev inequality.

6.3   RESPONSE TO RANDOM EXCITATION

In Section 6.2, where the random processes have been briefly discussed, we have shown that a random process may be described completely by a hierarchy of probability density functions or partially by a few moment functions. Once the probability information about a random process is available, the analysis is confined to obtaining the probability information about the response of a physical system. Schematically, this is represented as in Fig. 6.8. Mathematically,

$$L\bar{X}(t) = \bar{F}(t), \tag{6.49}$$

where L is an algebraic, differential or an integral operator defining a mathematical model of the given physical system. Given the probability information of $\bar{F}(t)$, the problem is to obtain the probability information about $\bar{X}(t)$.

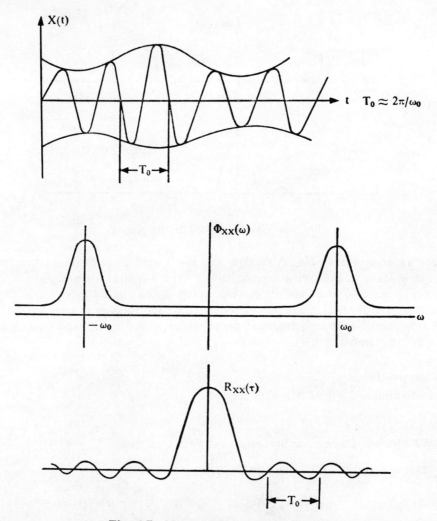

Fig. 6.7   Narrow Band Random Process.

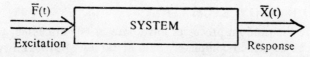

Fig. 6.8   Schematic Representation of Physical System.

We shall confine our discussion to systems modelled by linear differential operators with deterministic coefficients.

Example 6.1   Consider a mass-spring-dashpot system, as shown in Fig. 6.9. The equation of motion of the system is given by

$$Lx(t) = mf(t), \tag{6.50}$$

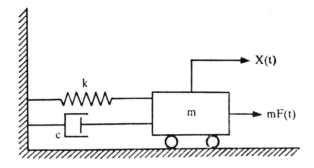

Fig. 6.9   Single-Degree-of-Freedom System.

where $x(t)$ and $f(t)$ are the state variables for the random processes $X(t)$ and $F(t)$ and

$$L = m\frac{d^2}{dt^2} + c\frac{d}{dt} + k.$$

Rewriting Eq. (6.50), we get

$$Lx = f(t), \qquad (6.51)$$

where

$$L = \frac{d^2}{dt^2} + 2\zeta\omega_0\frac{d}{dt} + \omega_0^2$$

with $\omega_0^2 = k/m$ and $\zeta = c/(2\omega_0 m)$.

Given

   excitation: $\mu_F(t)$, $K_{FF}(t_1, t_2)$;

   initial conditions: $x(t_0) = x_0$, $\dot{x}(t_0) = \dot{x}_0$.

To find

   response: $\mu_X(t)$, $K_{XX}(t_1, t_2)$.

   The general solution of Eq. (6.51) is given by

$$x(t) = g(t - t_0)x_0 + h(t - t_0)\dot{x}_0 + \int_{t_0}^{t} h(t - \tau)f(\tau)\,d\tau, \qquad (6.52)$$

where

   $Lg = 0$, $g(0) = 1, \dot{g}(0) = 0$;

   $Lh = 0$, $h(0) = 0, \dot{h}(0) = 1$;

$$g(t) = e^{-\zeta\omega_0 t}(\cos \omega_d t + \frac{p\omega_0 x_0}{\omega_d} \sin \omega_d t); \qquad (6.53)$$

$$h(t) = \frac{e^{-\zeta\omega_0 t}}{\omega_d} \sin \omega_d t; \qquad (6.54)$$

$$\omega_d = \omega_0\sqrt{1 - \zeta^2}. \qquad (6.55)$$

Since the integration and expectation operators commute,

$$\mu_X(t) = E[X(t)] = g(t - t_0)x_0 + h(t - t_0)\dot{x}_0 + \int_{t_0}^{t} h(t - \tau)\mu_F(\tau)\,d\tau, \qquad (6.56)$$

$$x(t_1) - \mu_X(t_1) = \int_{t_0}^{t_1} h(t_1 - \tau_1)[f(\tau) - \mu_F(\tau_1)]\, d\tau_1,$$

$$K_{XX}(t_1, t_2) = E[\{x(t_1) - \mu_X(t_1)\}\{x(t_2) - \mu_X(t_2)\}]$$

$$= \int_{t_0}^{t_2} \int_{t_0}^{t_1} h(t_1 - \tau_1) h^*(t_2 - \tau_2) E[\{f(\tau_1) - \mu_F(\tau_1)\}\{f(\tau_2) - \mu_F(\tau_2)\}]\, d\tau_1\, d\tau_2$$

$$= \int_{t_0}^{t_2} \int_{t_0}^{t_1} h(t_1 - \tau_1) h^*(t_2 - \tau_2) K_{FF}(\tau_1, \tau_2)\, d\tau_1\, d\tau_2. \tag{6.57}$$

$\mu_X(t)$ and $K_{XX}(t_1, t_2)$ can now be obtained from Eqs. (6.56) and (6.57).

Assertion

If $F(t)$ is stationary, i.e.,

$$\mu_F(t) = \mu_F = \text{constant},$$

then

$$K_{FF}(t_1, t_2) = R_{FF}(t_2 - t_1) \quad \text{for every } t_1 \text{ and } t_2$$

and, as $t_0 \to -\infty$ (or $t \to \infty$), $X(t)$ also is stationary.

If, in Example 6.1, $F(t)$ is stationary and

$$\mu_F = 0, \qquad R_{FF}(t_2 - t_1) = 2\pi\Phi_0\delta(t_2 - t_1),$$

and the initial conditions are

$$X(0) = \dot{X}(0) = 0,$$

the solution can be obtained as follows. Since $\mu_F = 0$, we have, from (6.56), $\mu_X = 0$. Substituting for $K_{FF}(t_1, t_2) = 2\pi\Phi_0\delta(t_2 - t_1)$ in (6.57), we have

$$R_{XX}(t_2 - t_1) = \int_0^{t_1} \int_0^{t_2} h(t_1 - \tau_1) h^*(t_2 - \tau_2) 2\pi\Phi_0\delta(\tau_2 - \tau_1)\, d\tau_1\, d\tau_2$$

$$= 2\pi\Phi_0 \int_0^{t_1} h(t_1 - \tau_1) h^*(t_2 - \tau_1)\, d\tau_1 \qquad (t_1 \leqslant t_2). \tag{6.58}$$

When $t_1 = t_2 = t$,

$$R_{XX}(0) = E[X^2(t)] = 2\pi\Phi_0 \int_0^t h^2(t - \tau)\, d\tau$$

$$= \frac{2\pi\Phi_0}{\omega_d^2} \int_0^t e^{-2\omega_0\zeta(t-\tau)} \sin^2 \omega_d(t - \tau)\, d\tau \quad \text{[substituting for } h(t-\tau) \text{ from Eq. (6.54)].} \tag{6.59}$$

Setting $t - \tau = \alpha$, we obtain

$$R_{XX}(0) = \sigma_X^2(t) = \frac{\pi\Phi_0}{\omega_d^2} \int_0^\infty e^{-2\omega_0\zeta\alpha}(1 - \cos 2\omega_d\alpha)\, d\alpha. \tag{6.60}$$

Evaluating the integral and taking the limit as $t \to \infty$ ($\alpha \to \infty$), we get

$$\sigma_X^2 = \frac{\pi\Phi_0}{\omega_d^2}\left(\frac{1}{2\omega_0\zeta} - \frac{\zeta}{2\omega_0}\right), \qquad \sigma_X = \left(\frac{\pi\Phi_0}{2\omega_0^3\zeta}\right)^{1/2}. \tag{6.61}$$

For the deterministic excitation $f(t) = A \sin pt$, we have seen that the amplitude of $X(t)$ is $X \approx 1/\zeta$. From Eq. (6.60) it is obvious that the rms value of $X(t)$ is $\sigma_X \approx 1/\sqrt{\zeta}$. Since $\zeta < 1$, damping is evidently more effective in reducing the response for random excitation. By differentiating Eq. (6.52) to get $\dot{x}(t)$ and carrying out the foregoing analysis, it can be shown that

$$R_{X\dot{X}}(0) = E[X\dot{X}] = 0,\tag{6.62}$$

$$R_{\dot{X}\dot{X}}(0) = E[\dot{X}^2] = \frac{\pi\Phi_0}{2\zeta\omega_0},\tag{6.63}$$

$$R_{\ddot{Y}\ddot{Y}}(0) = E[\ddot{Y}^2] = \frac{\pi\Phi_0}{2\zeta\omega_0}(1 + 4\zeta^2),\tag{6.64}$$

where $\ddot{Y} = \ddot{X} + F(t)$ = absolute acceleration. It should be noted that the nonstationary response of the simple oscillator can be obtained by evaluating the integral in (6.60).

The foregoing method of solution is called the *time domain approach.* We shall now discuss another method known as the *frequency domain approach.*

Frequency Domain Approach

The steady state response of a linear system may be expressed as

$$x(t) = \int_{-\infty}^{\infty} H(\omega)F(\omega)e^{i\omega t}\,d\omega,\tag{6.65}$$

where

$$F(\omega) = \frac{1}{\pi}\int_{-\infty}^{\infty} f(t)e^{-i\omega t}\,dt\tag{6.66}$$

and $H(\omega)$ is the frequency response function. The steady state response can be expressed also as

$$x(t) = \int_{-\infty}^{\infty} h(t - \tau)f(\tau)\,d\tau,\tag{6.67}$$

where $h(t - \tau)$ is the impulse response function. Setting $f(\tau) = \delta(\tau)$ in Eqs. (6.66) and (6.67), we get

$$F(\omega) = 1/(2\pi),\qquad x(t) = h(t).\tag{6.68}$$

Using Eq. (6.65), we obtain

$$h(t) = \frac{1}{2\pi}\int_{-\infty}^{\infty} H(\omega)e^{i\omega t}\,d\omega.\tag{6.69}$$

Hence, $h(t)$ is the inverse Fourier transform of $H(\omega)$. By the inversion theorem,

$$H(\omega) = \int_{-\infty}^{\infty} h(t)e^{-\omega t}\,dt.\tag{6.70}$$

Thus, $h(t)$ and $H(\omega)$ form the Fourier transform pair.

Computation of autocorrelation function   Let $F(t)$ be stationary and $\mu_F = 0$. Then (6.57) gives

$$R_{XX}(t_1, t_2) = \int_{t_0}^{t_1}\int_{t_0}^{t_2} h(t_1 - \tau_1)h^*(t_2 - \tau_2)R_{FF}(\tau_2 - \tau_1)\,d\tau_1\,d\tau_2.\tag{6.71}$$

Since $h(t - \tau) = 0$ for $\tau > t$, we have

$$R_{XX}(t_1, t_2) = \int_{t_0}^{\infty}\int_{t_0}^{\infty} h(t_1 - \tau_1)h^*(t_2 - \tau_2)R_{FF}(\tau_2 - \tau_1)\,d\tau_1\,d\tau_2.$$

Setting $t_2 - t_1 = \tau$, $t_1 - \tau_1 = u_1$, and $t_2 - \tau_2 = u_2$ and letting $t_0 \to -\infty$, we get

$$R_{XX}(\tau) = \int_{-\infty}^{\infty} \int_{-\infty}^{\infty} h(u_1) h(u_2) R_{FF}(\tau - u_1 - u_2) \, du_1 \, du_2. \tag{6.72}$$

Taking the Fourier transform of Eq. (6.72) and applying the convolution theorem of the Fourier integral, we obtain

$$\Phi_{XX}(\omega) = |H(\omega)|^2 \Phi_{FF}(\omega), \tag{6.73}$$

where $\Phi_{XX}(\omega)$, $\Phi_{FF}(\omega)$, and $H(\omega)$ are defined by Eqs. (6.73) and (6.70), respectively.

Thus, if the power spectral density (psd) $\Phi_{FF}(\omega)$ of the excitation is given, the psd of the response $\Phi_{XX}(\omega)$ can be computed from (6.73). Since the autocorrelation function and the psd form a transform pair, we have, by the inversion theorem,

$$R_{XX}(\tau) = \int_{-\infty}^{\infty} \Phi_{XX}(\omega) e^{i\omega\tau} \, d\omega, \tag{6.74}$$

$$\sigma_X^2 = E[X^2] = R_{XX}(0) = \int_{-\infty}^{\infty} \Phi_{XX}(\omega) \, d\omega = \int_{-\infty}^{\infty} |H(\omega)|^2 \Phi_{FF}(\omega) \, d\omega. \tag{6.75}$$

Approximate solution (for frequency domain technique)   A typical plot of $|H(\omega)|^2$ is shown in Fig. 6.10. For low values of damping, $|H(\omega)|^2$ is a very sharply-peaked function about a frequency ω_0. If the psd function is flat around ω_0, we may be justified in writing Eq. (6.75) as

$$\sigma_X^2 = \Phi_{FF}(\omega_0) \int_{-\infty}^{\infty} |H(\omega)|^2 \, d\omega \tag{6.76}$$

which effectively replaces the actual psd function by the constant psd of white noise. In such cases, we are fully justified in replacing a broad band excitation by white noise, even though

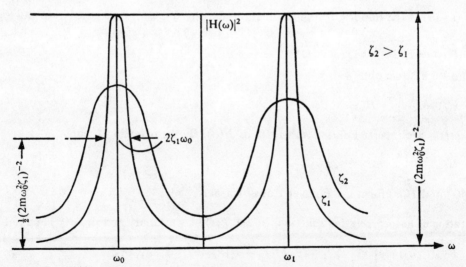

Fig. 6.10   Plots of Transmittancy Function.

the latter is not physically realizable. A typical example of acceptable and unacceptable white noise idealizations is shown in Fig. 6.11.

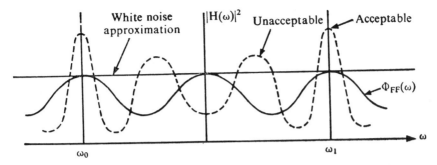

Fig. 6.11   Acceptable and Unacceptable White Noise Idealizations.

We now illustrate the frequency domain approach by considering Example 6.1. For the excitation

$$\mu_F = 0, \qquad \Phi_{FF}(\omega) = \Phi_0,$$

we have

$$H(\omega) = 1/(-\omega^2 - 2i\zeta\omega_0\omega + \omega_0^2),$$

$$|H(\omega)|^2 = 1/[(\omega_0^2 - \omega^2)^2 + 4\zeta^2\omega_0^2\omega^2].$$

Hence,

$$\Phi_{XX}(\omega) = \Phi_0/[(\omega_0^2 - \omega^2)^2 + 4\zeta^2\omega_0^2\omega^2].$$

From (6.75), we get

$$\sigma_X^2 = \Phi_0 \int_{-\infty}^{\infty} \frac{d\omega}{(\omega_0^2 - \omega^2)^2 + 4\zeta^2\omega_0^2\omega^2}. \tag{6.77}$$

The integrand in Eq. (6.77) has singularities in the domain of integration. The application of the Cauchy residue theorem in the complex domain reduces the integral to

$$\sigma_X^2 = \pi\Phi_0/(2\zeta\omega_0^3). \tag{6.78}$$

This equation is the same as Eq. (6.61) which we obtained by employing the time domain approach.

Results for structural damping   In many physical systems, it is realistic to assume structural damping. For a single-degree-of-freedom system, such as that considered in Example 6.1, the equation of motion with structural damping is given by

$$m\ddot{x} + k(1 + ig)x = mf(t). \tag{6.79}$$

It can be shown that, for Eq. (6.79),

$$|H(\omega)|^2 = \frac{1}{(\omega_0^2 - \omega^2)^2 + g^2\omega_0^4}, \tag{6.80a}$$

$$\sigma_X^2 = \pi\Phi_0/(g\omega_0^3). \tag{6.80b}$$

Application to design   When considering random excitation in structural design, the frequency domain approach is commonly used for two reasons: Firstly, it is relatively easy to obtain the psd by data processing; secondly, the psd provides valuable information about the frequency content of the excitation (this helps in choosing the natural frequency of the system to be so designed that the response is reduced).

Multidegree-of-Freedom Systems

Let $y(t)$ denote the displacement vector for a system with n degrees of freedom. The equation of motion of the system can be written as

$$[M]\ddot{y} + [C]\dot{y} + [K]y = p(t),\qquad(6.81)$$

where $[M]$, $[C]$ and $[K]$ are mass, damping, and stiffness matrices (each assumed to be real, symmetric, and positive definite) and $p(t)$ is the excitation vector.

Transformation to canonical form   Let us define a transformation $y = U_1 y_1$ such that

$$U_1^T M U_1 = M_1 = \text{diagonal},$$

$$U_1^T C U_1 = C_1,\qquad(6.82)$$

$$U_1^T K U_1 = K_1.$$

Then Eq (6.81) may be expressed as

$$M_1 \ddot{y}_1 + C_1 \dot{y}_1 + K_1 y_1 = U_1^T p(t).\qquad(6.83)$$

If we define $N = M_1^{-1/2}$ and set

$$y_1 = N y_2,\qquad(6.84)$$

and further let $N C_1 N = A$, $N C_2 N = B$, Eq. (6.83) becomes

$$I\ddot{y}_2 + A\dot{y}_2 + B y_2 = N U_1^T p(t),\qquad(6.85)$$

where A and B are real, symmetric, and positive definite. Under these conditions and the constraint $AB = BA$, it can be shown that there exists a transformation

$$y_2 = \theta x\qquad(6.86)$$

such that

$$\theta^T \theta = I,\qquad(6.87)$$

$$\theta^T A \theta = 2\omega_i \zeta_i,\qquad(6.88)$$

$$\theta^T B \theta = \omega_i^2.\qquad(6.89)$$

Under this transformation, Eq. (6.85) reduces to

$$\begin{bmatrix}\ddots & & \\ & 1 & \\ & & \ddots\end{bmatrix}\ddot{x} + \begin{bmatrix}\ddots & & \\ & 2\omega_i\zeta_i & \\ & & \ddots\end{bmatrix}\dot{x} + \begin{bmatrix}\ddots & & \\ & \omega_i^2 & \\ & & \ddots\end{bmatrix}x = \theta^T N U_1^T p(t) = f(t).\qquad(6.90)$$

In expanded form, Eq. (6.90) appears as

$$\ddot{x}_i + 2\omega_i\zeta_i\dot{x}_i + \omega_i^2 x_i^2 = f_i(t),\qquad i = 1, 2, \ldots, n,\qquad(6.91)$$

which is of the same form as (6.51). The probability characteristics of x_i ($i = 1, 2, \ldots, n$) can be obtained from (6.56) and (6.57).

Let $P(t)$ be a random excitation vector and the corresponding response be $Y(t)$, for which $p(t)$ and $y(t)$ are the corresponding state variable vectors. Working back through the transformation, we can obtain the probability information about $Y(t)$. Let

$$P(t) = aP(t), \tag{6.92}$$

where a is the deterministic constant vector. Setting $D = U_1N\theta$, we get

$$F(t) = D^T aP(t)$$

so that

$$F_j = (\sum_{k=1}^{n} d_k^j a_k)P(t) = \Gamma_j P(t), \tag{6.93}$$

where d^j is the j-th column of D.

To solve Eqs. (6.91), we can use the frequency domain approach to show that the covariant matrix of $Y(t)$ is

$$S_{YkYl}(0) = \sum_{i}^{n} \sum_{j}^{n} d_i^k d_j^l \Gamma_i \Gamma_j \int_{-\infty}^{\infty} |H_i(\omega)||H_j(\omega)|\Phi_{FF}(\omega)\, d\omega, \tag{6.94}$$

where $H_j(\omega)$ is the frequency response function of the j-th normal mode.

Approximate solution for response   It should be noted that, for light damping, $|H_i(\omega)|$ is sharply peaked about ω_i and, if ω_j and ω_k are well separated, $|H_i(\omega)||H_j(\omega)| \ll |H_j(\omega)|^2$. Also, since $d_i^k \Gamma_i$ may take both positive and negative values, $d_i^k d_j^l \Gamma_i \Gamma_j \ll (d_i^k \Gamma_i)^2$. Thus, we may write

$$.S_{YkYl} \approx \sum_{i}^{n} d_i^k d_i^l \Gamma_i^2 \int_{-\infty}^{+\infty} |H_i(\omega)|^2 \Phi_{FF}(\omega)\, d\omega. \tag{6.95}$$

By continuing the argument that $|H_i(\omega)|^2$ is sharply peaked about ω_i and assuming $\Phi_{FF}(\omega)$ to be locally white around a_i, approximation (6.95) may be written as

$$S_{YkYl} = \sum_{i}^{n} d_i^k d_i^l \Gamma_i^2 \frac{\Phi_{FF}(\omega_i)\pi}{2\zeta_i \omega_i^3}. \tag{6.96}$$

In particular, we get

$$S_{YkYk} = \sigma_{Yk}^2 = \sum_{i}^{n} (d_i^k)^2 \frac{\pi \Phi_{FF}(\omega_i)}{2\zeta_i \omega_i^3}. \tag{6.97}$$

Continuous systems   The differential equation of motion for a large class of structures is given by

$$Lw + \rho \frac{\partial^2 w}{\partial t^2} = F(x, t), \tag{6.98}$$

where $w(x, t)$ denotes the lateral displacement, and L is a linear differential operator representing the mathematical model of a structural element, e.g.,

$$L = T \frac{\partial^2}{\partial x^2} \qquad \text{(for string)},$$

$$L = \frac{\partial^2}{\partial x^2}\left(EI \frac{\partial^2}{\partial x^2}\right) \qquad \text{(for bending beam)},$$

$$L = \left(\frac{\partial^2}{\partial x_1^2} + \frac{\partial^2}{\partial x_2^2}\right)^2 \qquad \text{(for plate)}.$$

Let Eq. (6.98) be in one dimension and $u_n(x)$ represent the n-th mode shape. Then

$$Lu_n = \rho \omega_n^2 u_n. \tag{6.99}$$

Let

$$w(x, t) = \sum_{n=1}^{\infty} u_n y_n(t). \tag{6.100}$$

Substituting Eq. (6.100) in (6.98), we obtain

$$\sum_n (y_n L u_n + \rho u_n \frac{d^2 y_n}{dt^2}) = F(x, t).$$

Using Eq. (6.99), we have

$$\sum_n (\frac{d^2 y_n}{dt^2} + \omega_n^2 y_n)\rho u_n = F(x, t). \tag{6.101}$$

Multiplying both sides of Eq. (6.101) by u_m and using the orthogonality condition, we obtain

$$\int_0^1 \rho u_n u_m \, dx = 0 \qquad (m \neq n)$$
$$= M_n \qquad (m = n), \tag{6.102}$$

thereby reducing (6.101) to the canonical form

$$\frac{d^2 y_n}{dt^2} + \omega_n^2 y_n = \frac{1}{M_n} \int_0^1 u_n F(x, t) \, dx = g_n(t). \tag{6.103}$$

Assuming that dynamic coupling does not exist and introducing viscous damping, we may write

$$\frac{d^2 y_n}{dt^2} + 2\zeta_n \omega_n \frac{dy_n}{dt} + \omega_n^2 y_n = g_n(t) \tag{6.104}$$

which can be solved as in Section 6.3 and used in computing the probability information about $\omega(x, t)$.

In view of the similarity of Eqs. (6.104) and (6.91), the response of a continuous system to the random excitation $F(x, t)$ can be obtained. It should be observed that, in some applications, $F(x, t)$ may be a multiparametered random process.

Expected Frequency of Narrow Band Process

As already pointed out, the response of a resonant structure is a narrow band process with a narrow frequency spread around a resonant frequency. In such cases, it is worthwhile to calculate the expected value of the frequency and the expected number of crossings at a given value.

Let N_a^+ denote the number of crossings with a positive slope per unit time for a random process $X(t)$ at $x(t) = a$, as shown in Fig. 6.12. The time to cross da at a with a fixed velocity b is given by

$$\tau = da/|b|. \tag{6.105}$$

The probability of random process $X(t)$ lying in da at a with a velocity in db at b is given by $p_{X\dot{X}}(a, b) \, da \, db$. The expected number of crossings at $x(t) = a$ with $\dot{x}(t) = |b|$ per unit time is

$$p_{X\dot{X}}(a, b) \, da \, db/\tau = |b| p_{X\dot{X}}(a, b) \, db.$$

The expected number of crossings for any velocity at $x(t) = a$ is given by

$$N_a = \int_{-\infty}^{\infty} |b| p_{X\dot{X}}(a, b) \, db. \tag{6.106}$$

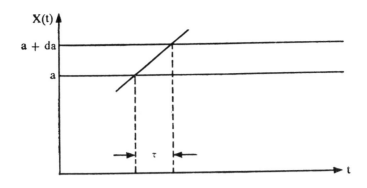

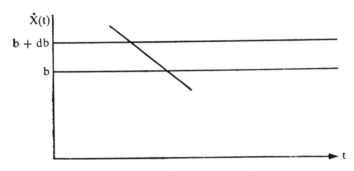

Fig. 6.12   Expected Frequency of Crossings.

The crossings with a positive slope for a narrow band random process are expressed as

$$N_a^+ = \tfrac{1}{2} \int_{-\infty}^{\infty} |b| p_{X\dot{X}}(a, b) \, db, \tag{6.107}$$

$$N_0^+ = \tfrac{1}{2} \int_{-\infty}^{\infty} |b| p_{X\dot{X}}(0, b) \, db. \tag{6.108}$$

If X and $\dot{X}$ are Gaussian with zero mean, we get

$$p_{X\dot{X}}(a, b) = \frac{1}{2\pi\sigma_X\sigma_{\dot{X}}} \exp \left[-\tfrac{1}{2}\left(\frac{a^2}{\sigma_X^2} + \frac{b^2}{\sigma_{\dot{X}}^2} \right) \right] \tag{6.109}$$

because X and $\dot{X}$ are independent. Substituting Eq. (6.109) in Eq. (6.107), we get

$$N_a^+ = \frac{1}{2\pi} \frac{\sigma_{\dot{X}}}{\sigma_X} \exp \left[-a^2/(2\sigma_X^2) \right], \tag{6.110}$$

$$N_0^+ = \frac{1}{2\pi} \frac{\sigma_{\dot{X}}}{\sigma_X}. \tag{6.111}$$

In Example 6.1, we obtained

$$\sigma_X = \left(\frac{\pi\Phi_0}{2\zeta\omega_0^3}\right)^{1/2}, \qquad \sigma_{\dot{X}} = \left(\frac{\pi\Phi_0}{2\zeta\omega_0}\right)^{1/2}.$$

Hence, $N_0^+ = \omega_0/(2\pi) = f_0$ which is the same as the natural frequency of the system and is independent of damping.

Probability distribution of peaks in narrow band process   The expected number of peaks per unit time occurring between a and $(a + da)$ is given by

$$P_a = N_a^+ - N_{a+da}^+ = -\frac{dN_a^+}{da}\, da. \tag{6.112}$$

The total number of peaks per unit time is N_0^+. Hence, the probability density function of peak distribution is

$$p_A(a) = -\frac{1}{N_0^+}\frac{dN_a^+}{da}. \tag{6.113}$$

For $X(t)$ and $\dot{X}(t)$, Gaussian N_a^+ and N_0^+ are given by Eqs. (6.110) and (6.111), respectively. Hence,

$$p_A(a) = \frac{a}{\sigma_X^2}\exp\left[-a^2/(2\sigma_X^2)\right] \tag{6.114}$$

which is a Rayleigh distribution.

BIBLIOGRAPHY

Bendat, J. S., Probability Functions for Random Processes: Prediction of peaks, fatigue damage, and catastrophic failure, NASA, CR–33, 1964.

Bendat, J. S., Enochson, L. D., Klein, G. H., and Piersol, A. G., The Application of Statistics to the Flight Vehicle Vibration Problem, ASDTR 61-123, Aeronautical Systems Division, AFSC, Wright Patterson AFB, Ohio, 1961.

Bendat, J. S., Enochson, L. D., Klein, G. H., and Piersol, A. G., Advanced Concepts of Stochastic Processes and Statistics for Flight Vehicle Vibration, Estimation and Measurement, ASDTDR 62–973, Aeronautical Systems Division, AFSC, Wright Patterson AFB, Ohio, 1962.

Bendat, J. S. and Piersol, A. G., Measurement and Analysis of Random Data, Wiley, New York, 1966.

Blackman, R. B. and Tukey, J. W., The Measurement of Power Spectra, Dover, New York, 1958.

Crandall, S. H., Random Vibration, Vols. I and II, MIT Press, Cambridge, Mass., 1958/1965.

Crandall, S. H. and Mark, W. D., Random Vibration, Academic Press, New York, 1963.

Davenport, W. B., Jr., and Root, W. L., Introduction to Random Signal and Noise, McGraw-Hill, New York, 1958.

Doob, J. L., Stochastic Processes, Wiley, New York, 1955.

Feller, W., An Introduction to Probability and Its Applications, Vols. I and II, Wiley, New York, 1957/1966.

Lin, Y. K., Probabilistic Theory of Structural Dynamics, McGraw-Hill, New York, 1967.

Papoulis, A., Probability, Random Variables and Stochastic Processes, McGraw-Hill, New York, 1965.

Parzen, W., Stochastic Processes, Holden Day, San Francisco, 1962.

Robson, J. D., An Introduction to Random Vibration, University Press, Edinburgh, 1963.

Wax, N., Selected Papers on Noise and Stochastic Processes, Dover, New York, 1954.

PART III OPTIMIZATION TECHNIQUES

7
Linear Programming

7.1 INTRODUCTION

As already stated in Section 1.1, decision-making is an important part of all human activity. In most problem situations, an individual is confronted with two or more solutions to the problem, one of which will give the best result. The selection of this key or optimal solution is based on a criterion called the *objective function*. A problem that deals with the maximization or the minimization of an objective function is said to be an *optimization problem*. We proceed to discuss the nature of an optimization problem.

Let A be any set whatsoever, B be any subset of real numbers, and $f : A \to B$, i.e., f is a rule that associates, to *each* element x in A, a *unique* element, to be denoted by $f(x)$, in B. In terms of this notation, an optimization problem may be stated as follows.

Find an element x_0 in A, called the optimal solution, if it exists, such that

$$f(x) \leqslant f(x_0) \quad \text{for all } x \text{ in } A \qquad (\textit{maximization problem})$$

or

$$f(x_0) \leqslant f(x) \quad \text{for all } x \text{ in } A \qquad (\textit{minimization problem}).$$

Depending on the nature of the basic parameters of an optimization problem, namely, f, A, and B, we distinguish between various types of optimization problems and thus form classes of such problems. Solution procedures are then developed for each class by considering its particular characteristics. Figure 7.1 gives a broad classification of optimization problems.

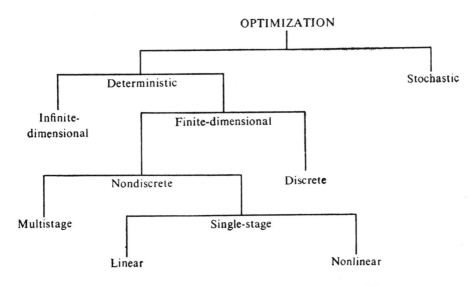

Fig. 7.1 **General Classification of Optimization Problems.**

For any optimization problem, the major questions that have so far been considered are:
  (i) Does an optimal solution exist?
  (ii) If an optimal solution exists, is it *unique*?
  (iii) Is there a *procedure* for finding an optimal solution, if it exists?
  (iv) How sensitive is the optimal solution to small changes in the parameters?

In the past few centuries, mathematicians have dealt successfully with some classes of optimization problems. But since 1940, several optimization problems that cannot be tackled by classical procedures have arisen. Some of these problems are classified as

  (i) linear programming;
  (ii) transportation;
  (iii) assignment;
  (iv) nonlinear programming;
  (v) dynamic programming;
  (vi) inventory;
  (vii) queuing;
  (viii) replacement;
  (ix) scheduling.

A novel feature of most of these problems is the presence of *inequality constraints*. Further, the solution procedure itself is *iterative*, i.e., given x^i in A, the solution procedure seeks to generate x^{i+1} in A, if it exists, so that $f(x^{i+1}) \geqslant f(x^i)$, assuming that we are maximizing the objective function. Another interesting feature is the *size* of the problem.

In this chapter, we shall consider the linear programming problem (LPP).

7.2 LINEAR PROGRAMMING PROBLEM

The linear programming problem is a very special case of the following particularization of the general optimization problem stated in Section 7.1. Let A be a subset of the n-dimensional Euclidean space, R^n. To be specific, the set A is given by

$$A = \{(x_1, x_2, \ldots, x_n) \mid g_i(x_1, x_2, \ldots, x_n) \leqslant b_i, i = 1, 2, \ldots, m\},$$

where, for $i = 1, 2, \ldots, m$, $g_i : X \to R$ (X being some subset of R^n) and b_i is a real number. This particularization of the general optimization problem is usually referred to as the *mathematical programming problem* (MPP).

The LPP is a particular case of the MPP just described. For an LPP, each of the $m + 1$ functions involved in the MPP formulation, i.e., g_i for $i = 1, 2, \ldots, m$, and f are assumed to be linear functions, and another restriction, namely, the non-negativity of the n variables x_j for $j = 1, 2, \ldots, n$, is usually imposed. Thus, the general LPP may be posed as follows.

Find a set of non-negative variables $x_1, x_2, \ldots, x_n$ satisfying the m constraints

$$\sum_{j=1}^{n} a_{ij} x_j (\leqslant \text{ or } \geqslant \text{ or } =) b_i \qquad (i = 1, 2, \ldots, m) \tag{7.1}$$

such that

$$\sum_{j=1}^{n} c_j x_j \tag{7.2}$$

is optimized (maximized or minimized). Restrictions (7.1) are called *constraints* of the LPP, and expression (7.2), which is to be optimized, is referred to as the *objective function*.

The essential characteristic of practical problems which can be formulated as LPPs may be described as follows. Consider some n activities which require m different resources. Let

$b_i =$ the available amount of the i-th resource;

$a_{ij} =$ the amount of the i-th resource required for one unit of the j-th activity;

$c_j =$ the profit corresponding to one unit of the j-th activity.

In addition to these parameters being known, suppose the following *linearity assumption* is valid.
If we indulge in x_j units of the j-th activity and x_k units of the k-th activity, then
(i) the amount of the i-th resource used for the j-th and k-th activities is

$a_{ij}x_j + a_{ik}x_k;$

(ii) the profit due to the j-th and k-th activities is

$c_j x_j + c_k x_k.$

Given this data, consider the following optimization problem. Find out the non-negative levels of each activity as permitted by the available resources so that the total profit is maximized. The mathematical model that solves such a problem is given as follows.
Find $x_1, x_2, \ldots, x_n$ such that

$$\sum_{j=1}^{n} a_{ij}x_j \leqslant b_i \qquad (i = 1, 2, \ldots, m),$$

$$x_j \geqslant 0 \qquad (j = 1, 2, \ldots, n) \tag{7.3}$$

and

$$\sum_{j=1}^{n} c_j x_j$$

is maximized. A problem of this type, which requires a linear objective function to be minimized (or maximized) subject to linear constraints, is called a *linear programming problem* (LPP).
For a better understanding of LPPs, let us consider a two-variable problem:

Maximize $2x + 7y$

such that

$$\begin{aligned} 2x + 3y &\leqslant 6, \\ 3x - 2y &\leqslant 6, \\ 0 \leqslant y &\leqslant 1, \\ x &\geqslant 0. \end{aligned} \tag{7.4}$$

It may be easily observed that (x, y) satisfies inequalities (7.4) if and only if it lies inside or on the polygon $OABCD$, as shown in Fig. 7.2. Our problem is to select the point (x, y), in the shaded region, which yields the maximum value of $2x + 7y$.

To solve the foregoing problem, we consider the parallel lines $2x + 7y = k$ for increasing values of k till we obtain a line that has at least one point in common with the shaded region. In our problem, we find that $2x + 7y = 10$ has only one point, namely, $C(\frac{3}{2}, 1)$, in common with the shaded region. Also, $2x + 7y = k$ for $k > 10$ does not have any point in common with the feasible region. Thus, we may conclude that the maximum value of $2x + 7y$ subject to the given restrictions is 10 and that this value is obtained when $x = 3/2$, $y = 1$. However, such a

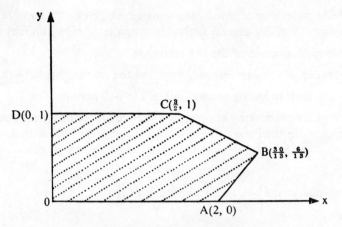

Fig. 7.2 Feasible Region.

solution procedure is not suitable when the number of variables is increased. Hence, we now consider a solution procedure that is applicable to any LPP.

7.3 SOLUTION PROCEDURE

Let the LPP be:

$$\text{Maximize } \sum_{j=1}^{n} c_j x_j$$

subject to

$$\sum_{j=1}^{n} a_{ij} x_j \leqslant b_i \qquad (i = 1, 2, \ldots, r), \tag{7.5}$$

$$\sum_{j=1}^{n} a_{ij} x_j \geqslant b_i \qquad (i = r + 1, \ldots, t), \tag{7.6}$$

$$\sum_{i=1}^{n} a_{ij} x_j = b_i \qquad (i = t + 1, \ldots, m), \tag{7.7}$$

$$x_j \geqslant 0 \qquad (j = 1, 2, \ldots, n), \tag{7.8}$$

where $b_i \geqslant 0$, $i = 1, 2, \ldots, m$ (assumed without any loss of generality).

For any solution $X = (x_1, x_2, \ldots, x_n)$ of constraints (7.5)–(7.8), we can find the non-negative real numbers $x_{n+1}, \ldots, x_{n+t}$ so that

$$\sum_{j=1}^{n} a_{ij} x_j + x_{n+i} = b_i \qquad (i = 1, 2, \ldots, r), \tag{7.9}$$

$$\sum_{j=1}^{n} a_{ij} x_j - x_{n+i} = b_i \qquad (i = r + 1, \ldots, t), \tag{7.10}$$

$$\sum_{j=1}^{n} a_{ij} x_j = b_i \qquad (i = t + 1, \ldots, m), \tag{7.11}$$

$$x_j \geqslant 0 \qquad (j = 1, 2, \ldots, n + t). \tag{7.12}$$

As introduced here, the variables $x_{n+1}, \ldots, x_{n+r}$ are called *slack variables*, and the variables x_{n+i} $(i = r + 1, \ldots, t)$ are known as *surplus variables*.

It can now be proved that maximizing

$$\sum_{j=1}^{n} c_j x_j$$

subject to (7.5)–(7.8) is equivalent to maximizing

$$\sum_{j=1}^{n} c_j x_j + \sum_{j=n+1}^{n+t} 0 x_j$$

subject to restrictions (7.9)–(7.12) in the following sense. Given an optimal solution $(x_1^0, x_2^0, \ldots, x_n^0)$ of the *original problem*, we can find the real numbers $x_{n+1}^0, \ldots, x_{n+t}^0$ so that $(x_1^0, \ldots, x_{n+t}^0)$ is an optimal solution of the *new problem*; conversely, for a given optimal solution $(x_1', x_2', \ldots, x_{n+t}')$ of the new problem, $(x_1', x_2', \ldots, x_n')$ is an optimal solution of the original problem.

It may be noted that the new problem has been obtained by replacing inequalities (7.5) and (7.6) by equalities (7.9) and (7.10) respectively. In doing this, the dimension of the original problem has increased. Such an increase is unavoidable because the mathematics of inequalities is not as straightforward as the mathematics of equalities. A major reason for this is that the inequalities satisfy

$$a \geqslant b \text{ and } c > 0 \quad \Rightarrow ac \geqslant bc,$$
$$a \geqslant b \text{ and } c < 0 \quad \Rightarrow ac \leqslant bc,$$

whereas the equalities satisfy

$$a = b \Leftrightarrow ac = bc \quad \text{for any real } c.$$

We can rewrite our problem in matrix form:

Maximize CX

subject to

$$AX = b, \qquad X \geqslant 0, \tag{7.13}$$

where

$$C = (c_1, \ldots, c_{n+t}), \qquad c_i = 0 \text{ for } i = n + 1, \ldots, n + t,$$
$$b = (b_1, \ldots, b_m),$$

and A is the coefficient matrix for the system of linear equations (7.9)–(7.11). Thus, our problem is to find out that non-negative solution of $AX = b$ for which CX achieves its maximum value.

It is obvious that if each component of C is negative and X^0 satisfies conditions (7.13) such that $CX^0 = 0$, then X^0 is an optimal solution of the LPP under consideration. The solution procedure given in this section is such that it reduces any solvable LPP to an equivalent LPP where an optimal solution of the type X^0 can be obtained. Before we describe this procedure, let us consider all the solutions of (7.13) and concentrate on a subset of these solutions that contains the optimal solution.

7.3.1   BASIC FEASIBLE SOLUTIONS

The system $AX = b$, $X \geqslant 0$, has (a) either no solution or just one solution or (b) an infinite number of solutions. The second alternative is obviously the only one worth considering. First, let us see how we can actually solve $AX = b$.

Assume that A has full row rank, i.e., the rows of A are linearly independent so that rank $(A) = m$. Then there exists a set of m linearly independent columns of A. Let us partition A as

$$A = (B \mid R),$$

where B contains any m linearly independent columns of A, and R has the remaining $n + t - m$ columns of A. Then $AX = b$ may be written as

$$[B \mid R]\begin{bmatrix} X_B \\ ---- \\ X_R \end{bmatrix} = b,$$

where X_B contains m components of X corresponding to the columns in B, and X_R has the remaining components of X corresponding to the columns in R. Thus, we have $BX_B = b - RX_R$ and, since B is nonsingular, we get $X_B = B^{-1}(b - RX_R)$.

Considering the different values of X_R, we can generate all the possible solutions of $AX = b$. For $X_R = 0$, we obtain $X_B = B^{-1}b$. The solution $(B^{-1}b, 0)$ of $AX = b$ is called a *basic solution*; if $(B^{-1}b, 0) \geqslant 0$, it is known as a *basic feasible solution* of $AX = b$, $X \geqslant 0$.

By considering different nonsingular matrices, e.g., B (called the basic matrix), we can generate all the basic solutions of $AX = b$. Some of these basic solutions are likely to satisfy also the non-negativity restrictions. Obviously, such solutions are finite in number.

We shall now give a geometric interpretation of the foregoing discussion. The solutions of $AX = b$, $X \geqslant 0$, are viewed as points in $(n + t)$-dimensional space. All the solutions give rise to a *convex polyhedron*, and the basic feasible solutions have a one-to-one correspondence with the vertices of this polyhedron. The importance of basic feasible solutions in linear programming becomes clear in the following theorem.

Fundamental theorem of LPP   *If an LPP has an optimal solution, then at least one basic feasible solution must be optimal.*

Thus, we have to concentrate on only the finite number of basic feasible solutions.

7.3.2 SIMPLEX ALGORITHM

The simplex algorithm provides a rule to find a basic feasible solution X^1, starting from a given basic feasible solution X^0, so that $CX^1 \geqslant CX^0$. This is done by replacing one column in B by another column in R, corresponding to X^0. The algorithm, at every iteration, indicates

(i) whether or not an optimal solution has been obtained;

(ii) the existence of an unbounded solution, i.e., the existence of a feasible solution X so that any large value of CX can be realized.

If the original problem has no feasible solution, the simplex algorithm detects this too. In short, the simplex algorithm is an *iterative procedure* for solving an LPP. The procedure leads to the optimal basic feasible solution, if it exists, in a *finite number of iterations*.

The only difficulty that remains is to obtain the initial basic feasible solution. But, fortunately, the iterative procedure can be applied to a *fictitious problem*, where the initial basic feasible solution is known in advance. After $m - r$ iterations, we would get a basic feasible solution of the original problem, if one exists. The fictitious problem associated with our problem is:

$$\text{Maximize} \quad \sum_{j=1}^{n+t+m-r} c_j x_j$$

subject to

$$\sum_{j=1}^{n} a_{ij}x_j + x_{n+i} = b_i \qquad (i = 1, 2, \ldots, r), \qquad (7.14)$$

$$\sum_{j=1}^{n} a_{ij}x_j - x_{n+i} + x_{n+t+i} = b_i \qquad (i = r+1, \ldots, t), \qquad (7.15)$$

$$\sum_{j=1}^{n} a_{ij}x_j + x_{n+t+i} = b_i \qquad (i = t+1, \ldots, m), \qquad (7.16)$$

$$x_j \geqslant 0 \qquad (j = 1, \ldots, n+t+m-r), \qquad (7.17)$$

where

$$c_j = 0 \qquad (j = n+1, \ldots, n+t), \qquad (7.18)$$

$$c_j = -M \qquad (j = n+t+1, \ldots, n+t+m-r), \qquad (7.19)$$

M being a very large positive number, usually about a thousand times the largest numerical value of c_j ($j = 1, 2, \ldots, n$). The variables x_j ($j = n+t+1, \ldots, n+t+m-r$) are called *artificial variables*.

The system of equations (7.14)–(7.16) may be represented in matrix notation as

$$\begin{bmatrix} A^1 & 0_{r,\,t-r} & I_r & 0_{r,\,t-r} & 0_{r,\,t-r} \\ A^2 & -I_{t-r} & 0_{t-r,\,r} & I_{t-r} & 0_{t-r,\,r} \\ A^3 & 0_{m-t,\,t-r} & 0_{m-t,\,r} & 0_{n-t,\,t-r} & I_{m-t} \end{bmatrix} \begin{bmatrix} X_{\text{original}} \\ X_{\text{surplus}} \\ X_{\text{slack}} \\ X_{\text{artificial}} \end{bmatrix} = b, \qquad (7.20)$$

where

$$A^1 = (a_{ij}), \qquad i = 1, \ldots, r, j = 1, 2, \ldots, n;$$
$$A^2 = (a_{ij}), \qquad i = r+1, \ldots, t, j = 1, 2, \ldots, n;$$
$$A^3 = (a_{ij}), \qquad i = t+1, \ldots, m, j = 1, 2, \ldots, n;$$

I_p = identity matrix of order p;

$0_{p,\,q}$ = null matrix of order $p \times q$;

$$X_{\text{original}} = \begin{bmatrix} x_1 \\ \vdots \\ x_n \end{bmatrix}; \quad X_{\text{surplus}} = \begin{bmatrix} x_{n+r+1} \\ \vdots \\ x_{n+t} \end{bmatrix}; \quad X_{\text{slack}} = \begin{bmatrix} x_{n+1} \\ \vdots \\ x_{n+r} \end{bmatrix}; \quad X_{\text{artificial}} = \begin{bmatrix} x_{n+t+1} \\ \vdots \\ x_{n+t+m-r} \end{bmatrix};$$

$$b = \begin{bmatrix} b_1 \\ \vdots \\ b_m \end{bmatrix}.$$

Thus, the fictitious LPP associated with the original LPP is of the following form:

Maximize CX

subject to

$$AX = b, \qquad X \geqslant 0, \qquad (7.21)$$

where $b \geqslant 0$ and the last m columns of A form the identity matrix I_m which gives us the initial basis matrix.

Computational Procedure

We now describe the sequential steps to be followed in solving an LPP.

Step 1   Ensure that every $b_i \geqslant 0$ $(i = 1, 2, \ldots, m)$; if $b_k < 0$, then multiply the k-th constraint by -1.

Step 2   Introduce slack, surplus, and artificial variables, if necessary, to obtain the fictitious LPP. Write the constraints as in Eq. (7.20). Let the fictitious LPP be:

$$\text{Maximize } CX$$

subject to

$$AX = b, \qquad X \geqslant 0,$$

where A is an $m \times p$ matrix. The last $m - r = q$ columns of A are the vectors associated with the artificial variables.

Step 3   Define a matrix

$$Y = (y_{ij}), \qquad i = 1, 2, \ldots, m + 1, j = 0, 1, \ldots, p,$$

where

$$y_{i0} = b_i \qquad\qquad (i = 1, 2, \ldots, m),$$

$$y_{m+1, 0} = -M \sum_{i=r+1}^{m} b_i,$$

$$y_{ij} = a_{ij} \qquad\qquad (i = 1, 2, \ldots, m, j = 1, 2, \ldots, p),$$

$$y_{m+1, j} = -M \sum_{i=r+1}^{m} y_{ij} - c_j \qquad (j = 1, 2, \ldots, p).$$

Here, y_{i0} gives the value of the i-th basic variable. Initially, the last m variables are the basic variables.

Step 4   Compute

$$\min_{1 \leqslant j \leqslant p} (y_{m+1, j}) = y_{m+1, k},$$

i.e., this minimum is achieved at $j = k$. If this minimum is attained at two or more indices, then select any one of them.

Step 5   If $y_{m+1, k} \geqslant 0$, the *optimal solution* has been obtained, so go to step 10. Otherwise, proceed to step 6.

Step 6   If $y_{m+1, k} < 0$, and $y_{ik} \leqslant 0$ for $i = 1, 2, \ldots, m$, then an *unbounded solution* exists. If this is not true, go to step 7.

Step 7   If $y_{m+1, k} < 0$ and $y_{ik} > 0$ for at least one i $(= 1, 2, \ldots, m)$, then compute

$$\min_{i \in I} \{y_{i0}/y_{ik}\},$$

where $I = \{i \mid y_{ik} > 0\}$, i.e., the minimum is to be computed over only those indices i where $y_{ik} > 0$. Let

$$\min_{i \in I} \{y_{i0}/y_{ik}\} = y_{r0}/y_{rk},$$

i.e., the minimum is attained for the index $i = r$. If there are many such indices where this minimum is achieved, then follow the procedure now described to determine r.

Compute

$$\min_{i \in I_p} \{y_{ip}/y_{ik}\}, \qquad p = 1, 2, \ldots,$$

where I_p is the set of indices at which

$$\min_{i \in I_{p-1}} \{y_{i, p-1}/y_{ik}\}$$

is achieved.

Step 8   Use k determined in step 4 and r in step 7 to define a new matrix

$$\hat{Y} = (\hat{y}_{ij}), \qquad i = 1, 2, \ldots, m + 1, j = 0, 1, \ldots, p,$$

where

$$\hat{y}_{ij} = y_{ij} - (y_{ik}y_{ij}/y_{rk}), \qquad i = 1, 2, \ldots, m + 1, i \neq r, j = 0, 1, \ldots, p;$$

$$\hat{y}_{rj} = y_{rj}/y_{rk}, \qquad\qquad j = 0, 1, 2, \ldots, p.$$

Step 9   After replacing y_{ij} by $\hat{y}_{ij}$, go to step 4. Now, $\hat{y}_{r0} = x_k$.

Step 10 y_{i0} gives the values of the optimal basic variables. All the nonbasic variables are put equal to zero. Also, we know y_{i0} in terms of x_j. In fact, if $y_{i0} = x_j$, then the j-th column of the Y-matrix must be the i-th unit vector.

To detect the possibility of the original problem having no feasible solution, we proceed as follows. Let $x_1, x_2, \ldots, x_n$ be the original variables, $x_{n+1}, \ldots, x_{n+t}$ the slack and surplus variables, and $x_{n+t+1}, \ldots, x_p$ the artificial variables. Then

(i) if none of the x_j's (> 0) associated with the unit vectors in step 10 is an artificial variable, $(x_1, x_2, \ldots, x_n)$, as computed in step 10, is an optimal solution of the original LPP; also, the maximum value of the objective function is given by $y_{m+1, 0}$;

(ii) if some of the x_j's (> 0) associated with the unit vectors in step 10 are artificial variables for any large value of M, the original LPP has no feasible solution.

A flowchart for the foregoing simplex algorithm is given in Fig. 7.3.

Example 7.1   We now illustrate the simplex procedure by solving a numerical problem:

Maximize $2x_1 + 5x_2 - 4.5x_3 + 1.5x_4$

subject to

$$5x_1 + 3x_2 + 1.5x_3 \leqslant 8,$$

$$1.8x_1 - 6x_2 + 4x_3 + x_4 \geqslant 3,$$

$$-3.6x_1 + 8.2x_2 + 7.5x_3 + 5x_4 = 15,$$

$$x_1 \geqslant 0, x_2 \geqslant 0, x_3 \geqslant 0, x_4 \geqslant 0.$$

Step 1   Since every $b_i \geqslant 0$, this step is over.

Step 2   Here, one slack, one surplus, and two artificial variables have to be introduced. Let

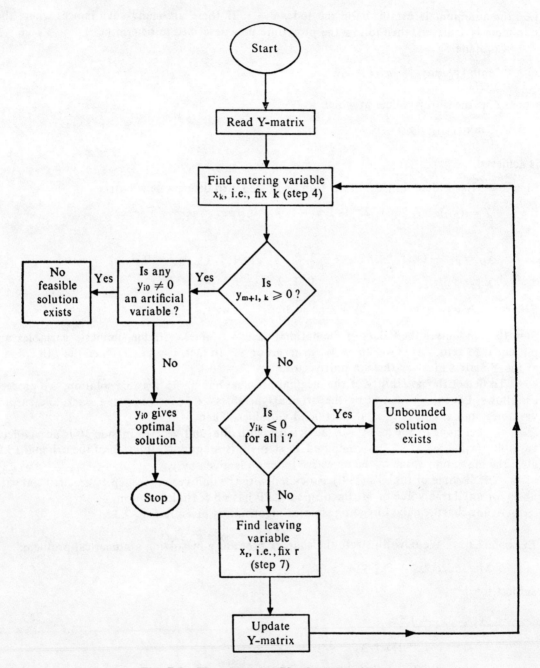

Fig. 7.3   Flowchart for Simplex Algorithm.

x_5 be the surplus variable, x_6 the slack variable, and x_7 and x_8 the artificial variables. Then the fictitious LPP is:

$$\text{Maximize } 2x_1 + 5x_2 - 4.5x_3 + 1.5x_4 + 0x_5 + 0x_6 - Mx_7 - Mx_8$$

subject to

$$\begin{bmatrix} 5 & 3 & 1.5 & 0 & 0 & 1 & 0 & 0 \\ 1.8 & -6 & 4 & 1 & -1 & 0 & 1 & 0 \\ -3.6 & 8.2 & 7.5 & 5 & 0 & 0 & 0 & 1 \end{bmatrix} \begin{bmatrix} x_1 \\ x_2 \\ \vdots \\ x_8 \end{bmatrix} = \begin{bmatrix} 8 \\ 3 \\ 15 \end{bmatrix}$$

with $x_j \geqslant 0 \ (j = 1, 2, \ldots, 8)$.

Step 3   In this step, for $M = 100$, the matrix Y is

x_6	8	5	3	1.5	0	0	1	0	0
x_7	3	1.8	−6	4	1	−1	0	1	0
x_8	15	−3.6	8.2	7.5	5	0	0	0	1
	−1800	178	−225	−1145.5	−601.5	100	0	0	0

Step 4   Since $p = 8$ and $m = 3$, it follows that

$$\min_{1 \leqslant j \leqslant 8} y_{4, j} = y_{4, 3} = -1145.5, \quad \text{i.e.,} \quad k = 3.$$

To indicate $k = 3$, we have marked the relevant column in step 3.

Step 5   Since

$$\min_{1 \leqslant j < 8} (y_{4, k}) < 0,$$

the optimal solution has not been obtained and so we go to step 6.

Step 6   Since $y_{i3} > 0$ for $i = 1, 2, 3$, the unbounded solution is not indicated.

Step 7   Since $y_{i3} > 0$ for $i = 1, 2, 3$, it follows that $I = \{1, 2, 3\}$. Also,

$$\min_{i \in I} \{y_{i0}/y_{i3}\} = \min \{\frac{8}{1.5}, \frac{3}{4}, \frac{15}{7.5}\} = \frac{3}{4} = \frac{y_{2,0}}{y_{2,3}}, \quad \text{i.e.,} \quad r = 2.$$

Step 8   Using the transformation given in step 8 of our computational procedure, we get the matrix $\hat{Y}$ as

x_6	6.88	4.32	5.25	0	−0.38	0.38	1	−0.38	0
x_3	0.75	0.45	−1.50	1	0.25	0.25	0	0.25	0
x_8	9.38	−6.97	19.45	0	3.13	1.88	0	−1.88	1
	−940.88	693.47	−1943.25	0	−315.13	−186.38	0	286.38	0

Step 9   Repeating the iterations from steps 4 to 8 of our computational procedure, we obtain the matrices (the various calculations at intermediate steps are indicated by the marked columns

and rows as

x_6	4.34	6.21	0	0	−1.22	−0.13	1	0.13	−0.27
x_3	1.47	−0.09	0	1	0.49	−0.11	0	0.11	0.08
x_2	0.48	−0.36	1	0	0.16	0.10	0	−0.10	0.05
	−4.22	−3.40	0	0	−2.91	0.96	0	99.04	99.91

x_1	0.70	1	0	1	−0.20	0.16	−0.02	0.02	−0.04
x_3	1.53	0	0	1	0.47	0.01	−0.11	0.11	0.07
x_2	0.73	0	1	0	0.09	0.06	0.09	−0.09	0.04
	−1.84	0	0	0	−3.57	0.55	0.88	99.12	99.76

x_1	1.34	1	0	0.41	0	0.17	−0.07	0.07	−0.01
x_4	3.24	0	0	2.11	1	0.03	−0.23	0.23	0.15
x_2	0.44	0	1	−0.19	0	0.06	0.11	−0.11	0.02
	9.73	0	0	7.54	0	0.65	0.08	99.92	100.32

Step 10   An optimal solution is ensured and is given by

$$x_1 = y_{10} = 1.34,$$
$$x_4 = y_{20} = 3.24,$$
$$x_2 = y_{30} = 0.44.$$

Also, the optimum value of the objective function is 9.73 which is given by y_{40}.

7.4   SENSITIVITY ANALYSIS

The optimal solution of the LPP

maximize CX

subject to

$$AX = b, \qquad X \geqslant 0,$$

depends on the *parameters* C, b, and A. The parameters are prone to error because they have to be estimated in any problem. If, after finding the optimal solution of this LPP, we are required to change *any one* of the components of any of these parameters, then the optimal solution of the altered problem can be easily computed by using the optimal solution of the original problem, particularly if the changes in the parameters are within certain limits. We now describe and illustrate these limits.

Changes in Vector C

Case I   Let c_j be such that x_j is not a basic variable. If this c_j is replaced by *any* $c_j + \Delta c_j$, where $-\infty < \Delta c_j < y_{m+1, j}$, then the optimal solution and the optimum value of the objective function do not change.

Case II   Let c_j be such that x_j is a basic variable, and let $x_j = y_{r0}$. Then this c_j can be replaced by *any* $c_j + \Delta c_j$, where

$$\max_{\substack{j \\ (y_{rj} > 0)}} \left(\frac{-y_{m+1, j}}{y_{rj}}\right) \leqslant \Delta c_j \leqslant \min_{\substack{j \\ (y_{rj} < 0)}} \left(\frac{-y_{m+1, j}}{y_{rj}}\right), \tag{7.22}$$

so that the optimal solution does not change. When Δc_j satisfies inequality (7.22), the optimum value of the objective function is $y_{m0} + \Delta c_j y_{r0}$.

Changes in Vector b

If the l-th component b_l of the vector b is replaced by $b_l + \Delta b_l$, where

$$\max_{\substack{i \\ (y_{i, p-m+l} > 0)}} \left\{\frac{-y_{i0}}{y_{i, p-m+l}}\right\} \leqslant \Delta b_l \leqslant \min_{\substack{i \\ (y_{i, p-m+l} < 0)}} \left\{\frac{-y_{i0}}{y_{i, p-m+l}}\right\}, \tag{7.23}$$

then no more iterations of the simplex algorithm are required. Further, if, in the current Y-matrix, we replace y_{i0} by

$$\hat{y}_{i0} = y_{i0} + \Delta b_l y_{i, p-m+l} \qquad (i = 1, 2, \ldots, m),$$

the computations in step 10 give an optimal solution of the altered problem in which b_l is replaced by $b_l + \Delta b_l$, provided Δb_l satisfies inequality (7.23).

Changes in Matrix A

Let a_{ij} be replaced by $a_{ij} + \Delta a_{ij}$. We now give the limits for Δa_{ij} so that, with minor computations, the optimal solution for the changed problem can be obtained.

Case I   Let the j in a_{ij} be such that x_j is not a basic variable. Also, let

$$C_B = (C_{B1}, C_{B2}, \ldots, C_{Bm}),$$

where $C_{Bi} = C_j$ if $x_j = y_{i0}$. With this notation, if

$$\Delta a_{ij} \geqslant \frac{-y_{m+1, j}}{\sum\limits_{r=1}^{m} C_{Br} y_{r, p-m+i}} \qquad \left(\text{when } \sum_{r=1}^{m} C_{Br} y_{r, p-m+i} > 0\right) \tag{7.24}$$

or

$$\Delta a_{ij} \leqslant \frac{-y_{m+1, j}}{\sum\limits_{r=1}^{m} C_{Br} y_{r, p-m+i}} \qquad \left(\text{when } \sum_{r=1}^{m} C_{Br} y_{r, p-m+l} < 0\right), \tag{7.25}$$

neither the optimal solution nor the optimum value of the objective function changes when a_{ij} is replaced by any $a_{ij} + \Delta a_{ij}$, provided Δa_{ij} satisfies inequality (7.24) or inequality (7.25) whichever is the case. Also, if

$$\sum_{r=1}^{m} C_{Br} y_{r, p-m+i} = 0,$$

then this statement is valid for any value of Δa_{ij}.

Case II Let the j in a_{ij} be such that x_j is a basic variable, and let $x_j = y_{t0}$. If Δa_{ij} satisfies

$$\max_{k \neq t} \{\frac{-y_{k,\,0}}{A_k}\} \leqslant \Delta a_{ij} \leqslant \min_{k \neq t} \{\frac{-y_{k,\,0}}{A_k}\}, \qquad (7.26)$$

where

$$A_k = y_{k,\,0}y_{t,\,p-m+i} - y_{t,\,0}y_{k,\,p-m+i},$$

and

$$\max_{\substack{q \\ (B_q > 0)}} \{\frac{-y_{m+1,\,q}}{B_q}\} \leqslant \Delta a_{ij} \leqslant \min_{\substack{q \\ (B_q < 0)}} \{\frac{-y_{m+1,\,q}}{B_q}\}, \qquad (7.27)$$

where

$$B_q = y_{m+1,\,q}y_{t,\,p-m+i} - y_{iq} \sum_{r=1}^{m} C_{Br}y_{r,\,p-m+i}$$

(with $C_{B1}, C_{B2}, \ldots, C_{Bm}$ being the same as defined in case I), then the optimal solution of the changed problem can be obtained without any further iteration of the simplex algorithm. In fact, if y_{k0} for $k = 1, 2, \ldots, m$ in the current Y-matrix is replaced by

$$\hat{y}_{k0} = y_{k0} - \frac{\Delta a_{ij}y_{k,\,p-m+i}}{1 + \Delta a_{ij}y_{t,\,p-m+i}} \qquad (k = 1, 2, \ldots, m, k \neq t),$$

$$\hat{y}_{ti} = \frac{y_{t0}}{1 + \Delta a_{ij}y_{t,\,p-m+i}},$$

then, by using step 10, we can compute the optimal solution of the new problem when a_{ij} is replaced by $a_{ij} + \Delta a_{ij}$, provided Δa_{ij} satisfies inequality (7.26) as well as inequality (7.27). In other words, if $x_j = y_{t0}$, then $\hat{x}_j = \hat{y}_{t0}$, where $\hat{x}_j$ refers to the changed problem.

BIBLIOGRAPHY

Chun, An-Ming, Linear Programming, Merrill, Columbus, Ohio, 1963.

Dano, S., Linear Programming in Industry, 4th edn., Springer Verlag, New York, 1974.

Dantzig, G. B., Linear Programming and Extensions, Princeton University Press, Princeton, 1963.

Hadley, G., Linear Programming, Addison-Wesley, Reading, Mass., 1962.

Murthy, K. G., Linear and Combinatorial Programming, Wiley, New York, 1977.

Simmonard, M., Linear Programming, Prentice-Hall, Englewood Cliffs, New Jersey, 1966.

Zionts, S., Linear and Integer Programming, Prentice-Hall, Englewood Cliffs, New Jersey, 1974.

8

Nonlinear Programming

8.1 INTRODUCTION

The general nonlinear programming (NLP) problem may be stated as:

$$\text{Minimize } f(X) \qquad\qquad (8.1)$$

subject to

$$g_i(X) \leqslant 0 \qquad (i = 1, 2, \ldots, m), \qquad\qquad (8.2)$$

$$X \geqslant 0, \qquad\qquad (8.3)$$

where

$$X = (x_1, x_2, \ldots, x_n).$$

In minimum weight design problems, the objective function $f(X)$ gives the weight of the structure for the design X, and constraints (8.2) and (8.3) ensure that X is feasible; these constraints include both behaviour and side constraints.

In general, structural problems are described by equilibrium and compatibility equations which contain design parameters—such as area and second moment of area—and behaviour variables, for example, stress, strain, moment, deflection, and natural frequency. If the design parameters are to be determined also for optimality, then some of the equilibrium and compatibility equations are nonlinear. Since nonlinear constraints are unavoidable in elastic design, NLP problems often occur in structural optimization.

One of the procedures for solving an NLP problem entails approximating the problem functions f and g_i by the first-order term of their Taylor series expansion using a given X_0, and thereby obtaining a corresponding LPP. The solution of this LPP is used in place of X_0, yielding a sequence of LPPs, the solutions of which, under certain conditions, converge to the solution of the NLP problem. *Cutting plane methods*, for example, are of this type. It is important to note that theoretical considerations indicate that each 'well behaved' NLP problem is equivalent to 'some' LPP. *Linearization methods* attempt to obtain this LPP through some approximations.

Another type of technique—for example, *methods of feasible directions* and *of gradient projection*—tackles the NLP problem without converting it into an LPP. Some efficient algorithms are designed to solve problems in which all the constraints are linear. However, we shall not discuss such algorithms. We shall confine ourselves to the *sequential unconstrained minimization technique* (SUMT) which is essentially a generalization of the Lagrange method of undetermined multipliers. This class of methods, generally considered the most important, can be broadly subdivided into two categories: (a) *interior method* and (b) *exterior method*. In the interior formulation, all the unconstrained minima lie in the feasible region and converge to the solution for well behaved problems. This method is particularly suitable for structural applications. Its major advantage is that, at each iteration, a feasible design, which possesses some built-in safety measure to take care of both occasional overloads and some factors neglected at the analysis stage, is available. Our discussion centres on this technique.

8.2   ROLE OF CONVEXITY IN NLP

In NLP, the objective function f has to be optimized over the feasible region, say S. Most methods require that we solve the following type of problem several times:

$$\min_{\alpha} f(X_0 + \alpha d)$$

such that $X_0 + \alpha d \in S$, where X_0 and d are fixed. If the function f and the set S are not well behaved, then this problem too is very complex. Hence, we shall briefly review the conditions under which the problem becomes well behaved.

8.2.1   CONVEX SET

A set $S \subseteq R^n$ is said to be *convex* if, for any two X_1, X_2 in S, $\lambda X_1 + (1 - \lambda)X_2 \in S$ for $0 \leqslant \lambda \leqslant 1$. Geometrically, this implies that a set S is convex if every point on the line segment joining any two points in the set S lies in the set.

8.2.2   CONVEX-LIKE FUNCTIONS

An NLP problem requires that we find the *global minimum* of the objective function over the feasible region. However, since all the available methods compare the objective function values only in the vicinity of any given point, they can at best locate a *local minimum*. Therefore, it is very important to know the conditions which ensure that a local minimum is global. Interestingly enough, such conditions are quite often satisfied by physical problems. We shall now discuss these conditions.

Let R be the set of real numbers, S a convex subset of R^n, n-Euclidean space. Also, let $f : S \to R$ be a continuous function and C' be the class of all such continuous functions that possess all the first-order partial derivatives which, in turn, are continuous over the interior of S. Let

$$\nabla f(X) = (\frac{\partial f}{\partial x_1}, \frac{\partial f}{\partial x_2}, \ldots, \frac{\partial f}{\partial x_n})$$

denote the gradient of f at X. With this notation, a function f, defined over a convex set S, is said to be

(i) *strictly convex* if, for all $\lambda \in [0, 1]$,

$$f[\lambda X_1 + (1 - \lambda)X_2] < \lambda f(X_1) + (1 - \lambda)f(X_2)$$

or, equivalently, for $f \in C'$, if

$$f(X_1) - f(X_2) > (X_1 - X_2)' \nabla f(X_2);$$

(ii) *convex* if, for all $\lambda \in [0, 1]$,

$$f[\lambda X_1 + (1 - \lambda)X_2] \leqslant \lambda f(X_1) + (1 - \lambda)f(X_2)$$

or, equivalently, for $f \in C'$, if

$$f(X_1) - f(X_2) \geqslant (X_1 - X_2)' \nabla f(X_2);$$

(iii) *pseudoconvex* if

$$(X_1 - X_2)' \nabla f(X_2) \geqslant 0 \Rightarrow f(X_1) \geqslant f(X_2)$$

or, equivalently, if

$$f(X_1) < f(X_2) \Rightarrow (X_1 - X_2)' \nabla f(X_2) < 0;$$

(iv) *strictly quasiconvex* if, for all $\lambda \in [0, 1]$,

$$f(X_1) < f(X_2) \Rightarrow f[\lambda X_1 + (1 - \lambda)X_2] < f(X_2);$$

(v) *quasiconvex* if, for all $\lambda \in [0, 1]$,

$$f(X_1) \leqslant f(X_2) \Rightarrow f[\lambda X_1 + (1 - \lambda)X_2] \leqslant f(X_2)$$

or, equivalently, for $f \in C'$, if

$$(X_1 - X_2)' \nabla f(X_2) > 0 \Rightarrow f(X_1) > f(X_2)$$

or, equivalently, for $f \in C'$, if

$$f(X_1) \leqslant f(X_2) \Rightarrow (X_1 - X_2)' \nabla f(X_2) \leqslant 0.$$

In these definitions, we have assumed that X_1, X_2 are any two points of S and that $f \in C'$ whenever the derivatives are required.

Apart from some 'pathological' cases, the inclusion relations that can be proved are

strict convexity $\Rightarrow$ convexity $\Rightarrow$ pseudoconvexity $\Rightarrow$ strict quasiconvexity

$\Rightarrow$ quasiconvexity.

The results that account for the importance of these functions in NLP are now given.

(i) If S is a convex set, then every local minimum of the problem

$$\min_{X \in S} f(X)$$

is a global minimum if and only if the function f is strictly quasiconvex on *every* convex subset of S.

(ii) The set $A(\alpha) = \{X \in S \,|\, f(X) \leqslant \alpha\}$ is convex for *every* α if and only if the function f is quasiconvex.

Thus, if each g_i is quasiconvex in (8.2), then the constraint region is convex. Further, if the objective function (8.1) is strictly quasiconvex, then we need not distinguish between local minimum and global minimum.

We should note that it is relatively easier to test the convexity of a function than, for example, its pseudoconvexity. This accounts for the importance of smaller classes of functions.

A class of concave-like functions which corresponds to the foregoing class of convex-like functions is now briefly described. If $-f$ is a convex function, then f is called a concave function. This analogy is extended to all the five classes of convex-like functions. For concave-like functions, the minimum is replaced by the maximum in result (i) and $\leqslant$ is replaced by $\geqslant$ in result (ii). It may be observed that the geometrical interpretations of the conditions defining these classes of functions are very helpful in studying them.

We may also remark that most of the theoretically sound solution procedures are valid when f is strictly quasiconvex and each g_i is quasiconvex. If the functions do not satisfy these conditions, the procedure is still applied in the hope that it might be valid near the optimum point. This is true for most structural design problems where it is difficult to assess the convexity of the objective as well as the constraint functions.

8.3   KUHN-TUCKER CONDITIONS

A classical result of Lagrange converts an equality-constrained optimization problem into an equivalent unconstrained optimization problem. This, in turn, is equivalent to solving a system of nonlinear equations. For optimization problems in which the constraints are inequalities, Kuhn and Tucker have obtained certain nonlinear inequalities whose solution gives a solution to the optimization problem. We shall discuss these conditions in this section.

For the problem given by (8.1)–(8.3), we define the Lagrangian function

$$\phi(X, u) = f(X) + \sum_{i=1}^{m} u_i g_i(X). \tag{8.4}$$

The following results can now be proved.

(i) If any $(\hat{X}, \hat{u}) \geqslant 0$ satisfies

$$\phi(\hat{X}, u) \leqslant (\hat{X}, \hat{u}) \leqslant \phi(X, \hat{u}) \quad \text{for all } X \geqslant 0, \, u \geqslant 0, \tag{8.5}$$

then $\hat{X}$ solves the NLP problem defined by (8.1)–(8.3).

(ii) If $\hat{X} \geqslant 0$ solves the NLP problem given by (8.1)–(8.3), and each of the functions f, g_i is convex, then a $\hat{u} \geqslant 0$ can be found such that $(\hat{X}, \hat{u})$ satisfies inequality (8.5).

8.3.1   SADDLE POINT

A point $(\hat{X}, \hat{u})$ satisfying (8.5) is called a *saddle point* of $\phi(X, u)$.

In view of the foregoing results, it follows that a convex programme has a solution if and only if a saddle point of the Lagrangian function exists. Also, the X-part of any saddle point solves a given programme. Hence, solving an NLP problem may be regarded as equivalent to finding out a saddle point of the Lagrangian function.

8.3.2   KUHN-TUCKER CONDITIONS—DIFFERENTIAL FORM

If the function f is strictly quasiconvex and each of the functions g_i is pseudoconvex, then the following conditions are equivalent to the existence of a saddle point $(\hat{X}, \hat{u})$:

$$\left.\frac{\partial \phi}{\partial x_i}\right|_{(\hat{x}, \hat{u})} \geqslant 0 \qquad (i = 1, 2, \ldots, n), \tag{8.6}$$

$$\hat{X}_i \geqslant 0 \qquad (i = 1, 2, \ldots, n), \tag{8·7}$$

$$\sum_{i=1}^{n} \hat{X}_i \left.\frac{\partial \phi}{\partial x_i}\right|_{(\hat{x}, \hat{u})} = 0, \tag{8.8}$$

$$\left.\frac{\partial \phi}{\partial u_j}\right|_{(\hat{x}, \hat{u})} \leqslant 0 \qquad (j = 1, 2, \ldots, m), \tag{8.9}$$

$$\hat{u}_j \geqslant 0 \qquad (j = 1, 2, \ldots, m), \tag{8.10}$$

$$\sum_{j=1}^{m} \hat{u}_j \left.\frac{\partial \phi}{\partial u_j}\right|_{(\hat{x}, \hat{u})} = 0. \tag{8.11}$$

Conditions (8.6)–(8.11) are called Kuhn-Tucker conditions for the problem defined by (8.1)–(8.3). In other words, an NLP problem may be considered equivalent to a system of nonlinear equalities and inequalities. If, for a given NLP problem, (8.6)–(8.11) can be solved, then the X-part of the solution gives a solution of the NLP problem. It may be remarked that even if the problem functions do not satisfy the appropriate convexity requirements, the X-part of the solution of (8.6)–(8.11) can give a solution of the NLP problem.

The Kuhn-Tucker conditions can be further proved to be equivalent to the following conditions.

$\hat{X} \geqslant 0$ solves the NLP problem if and only if there exist $\hat{u}_i$ and $\hat{v}_i$ such that

$$-\nabla f(\hat{X}) = \sum_{j \in \hat{M}} \hat{u}_j \nabla g_j(\hat{X}) + \sum_{i \in \hat{N}} \hat{v}_i(-e_i), \tag{8.12}$$

$$\begin{aligned} \hat{u}_j &\geqslant 0 \quad (j \in \hat{M}), \\ \hat{v}_i &\geqslant 0 \quad (i \in \hat{N}), \end{aligned} \tag{8.13}$$

where

$$\hat{M} = \{j \mid g_j(\hat{X}) = 0\}, \tag{8.14}$$

$$\hat{N} = \{i \mid \hat{X}_i = 0\}, \tag{8.15}$$

and e_i is the i-th unit vector of n components.

The necessary and sufficient conditions (8.12)–(8.15), which characterize an optimal solution of the NLP problem, can be interpreted geometrically as follows. A point $\hat{X}$ solves an NLP problem if and only if the negative gradient of the objective function at $\hat{X}$ can be expressed as a non-negative linear combination of the exterior unit normals to the binding surfaces that pass through $\hat{X}$.

Using (8.12)–(8.15), we can show that the NLP problem given by (8.1)–(8.3) is equivalent to some LPP. To be precise, we can prove the following.

$\hat{X}$ solves an NLP problem if and only if $\hat{X}$ is a solution of the LPP:

$$\text{Minimize } X \nabla f(\hat{X}) \tag{8.16}$$

subject to

$$(X - \hat{X}) \nabla g_j(\hat{X}) \leqslant 0 \quad (j \in \hat{M}), \tag{8.17}$$

$$X_i \geqslant 0 \quad (i \in \hat{N}). \tag{8.18}$$

It should be noted that this LPP depends on the solution $\hat{X}$ of the NLP problem to be solved. For this reason, attempts have been made to obtain a sequence of the LPPs that converge to the LPP (8.16)–(8.18).

Although we cannot clearly show the role that the Kuhn-Tucker conditions have played in the development of various solution procedures, we are sure that any theoretically sound algorithm makes use of these conditions in some form. We now give an example to illustrate the fact that some structural optimization problems can be directly solved with the help of the Kuhn-Tucker conditions (8.6)–(8.11). (This example is discussed in greater detail in Chapter 11.)

Example 8.1

$$\text{Minimize } f(X) = x_1 x_2 (2 + x_3) \tag{8.19}$$

subject to

$$g_1(X) = P - \frac{k_1 x_1^3 x_2 x_3^3}{8 + 6x_3^2 + x_3^3} \leqslant 0, \tag{8.20}$$

$$g_2(X) = P - k_2 x_1^{0.25} x_2^{1.75} (2 + x_3) \leqslant 0, \tag{8.21}$$

$$x_1 > 0, \ x_2 \geqslant 0, \ x_3 \geqslant 0. \tag{8.22}$$

Here, $f(X)$ gives the area of cross-section, condition (8.20) is the primary buckling constraint for the channel-section column, and condition (8.21) is the crippling behaviour constraint for the Z-section column.

The Lagrangian function is

$$\phi(X, u) = f(X) + u_1 g_1(X) + u_2 g_2(X). \tag{8.23}$$

The Kuhn-Tucker conditions state that $\hat{X}$ solves problem (8.19)-(8.22) if, for $\phi(X, u)$ given by Eq. (8.23), conditions (8.6)-(8.11) are satisfied for some $\hat{u}$. In particular, (8.11) requires that

$$\hat{u}_1 g_1(\hat{X}) + \hat{u}_2 g_2(\hat{X}) = 0.$$

Consider the four possibilities

  (i) $g_1(\hat{X}) < 0, \ g_2(\hat{X}) < 0;$

  (ii) $g_1(\hat{X}) = 0, \ g_2(\hat{X}) < 0;$

  (iii) $g_1(\hat{X}) < 0, \ g_2(\hat{X}) = 0;$

  (iv) $g_1(\hat{X}) = 0, \ g_2(\hat{X}) = 0.$

It may be easily argued that, in (i), (ii), and (iii), the remaining Kuhn-Tucker conditions, i.e., (8.6)-(8.10), cannot be satisfied for any $(\hat{X}, \hat{u})$. Thus, only (iv) can satisfy these conditions. It can be verified that, in this case,

$$\hat{x}_1 = (2.02) P^{0.15} k_2^{0.2} k_1^{-0.35},$$

$$\hat{x}_2 = (0.752) P^{0.55} k_1^{0.05} k_2^{-0.6},$$

$$\hat{x}_3 = 1.68899,$$

$$\hat{u}_1 = \frac{0.3(2 + \hat{x}_3)(8 + 6\hat{x}_3^2 + \hat{x}_3^3)}{k_1 \hat{x}_1^2 \hat{x}_3^3},$$

$$\hat{u}_2 = \frac{0.4}{k_2} \left[\frac{\hat{x}_1 (2 + \hat{x}_3)}{\hat{x}_2} \right]^{0.75}$$

satisfy each of the Kuhn-Tucker conditions (8.6)-(8.11). Hence, $(\hat{x}_1, \hat{x}_2, \hat{x}_3) = \hat{X}$ gives a solution to the optimization problem (8.19)-(8.22). It should be noted that $\hat{x}_3$ is independent of the problem parameters.

8.4 LINEARIZATION TECHNIQUE

The linearization technique is based on the following property. If g is convex, then

$$\{X \mid g(X) \leqslant 0\} \subseteq \{X \mid g(X^0) + (X - X^0) \nabla g(X^0) \leqslant 0\},$$

where X^0 is any arbitrary point. Thus, the constraint region given by the nonlinear constraints can be enveloped by linear constraints which are obtained through the first-order terms of the Taylor series. Also, the problem defined by (8.1)-(8.3) is equivalent to:

Minimize x_{n+1}

subject to

$$g_i(X) \leqslant 0 \qquad (i = 1, 2, \ldots, m), \tag{8.24}$$

$$g_{m+1} = f(X) - x_{n+1} \leqslant 0,$$
$$X \geqslant 0.$$

(8.25)

This equivalent formulation shows that the objective function can always be taken to be linear. Problem (8.24)–(8.25) when coupled with the linearization property just referred to gives an LPP to be solved at each iteration.

8.4.1 CUTTING PLANE ALGORITHM

By applying the foregoing properties, we can state the cutting plane algorithm as follows.

Step 1   Initialize, i.e., choose any X^0, not necessarily feasible, and set $k = 0$.

Step 2   Solve the LPP:

Minimize x_{n+1}

subject to

$$g_i(X^k) + (X - X^0)\nabla g_i(X^k) \leqslant 0 \qquad (i = 1, 2, \ldots, m + 1),$$
$$X \geqslant 0.$$

Let the solution of the LPP be X^{k+1}.

Step 3   If X^{k+1} satisfies the nonlinear constraints (8.24), terminate. Otherwise, choose the most violated constraint, i.e.,

$$g_r(X^{k+1}) = \max_i g_i(X^{k+1});$$

this means the r-th constraint is the most violated one.

Step 4   Linearize g_r with respect to X^{k+1} and add it to the LPP in step 2.

Step 5   Solve the new LPP and return to step 3.

It may be remarked that an additional constraint is added in each successive LPP. To cope with such a situation effectively, the dual simplex algorithm may be used.

A major drawback of the cutting plane algorithm is that, at every stage, only an infeasible point is available. The algorithm converges for any arbitrary choice of X^0, but the convergence rate depends on this choice.

8.5 SINGLE-VARIABLE SEARCH

The methods considered in this section and the following one are used in Section 8.7. Let us consider the problem

$$\min_{x \in [a, b]} f(x).$$

This kind of a problem needs to be solved several times while solving an NLP problem by means of any procedure. We shall concentrate on those problems where the function f is quasi-convex (*unimodal*), i.e.,

$$a \leqslant x_1 < x_2 < x_3 < b \Rightarrow f(x_2) \leqslant \max \, [f(x_1), f(x_3)]$$

for any choice of x_1, x_2, x_3 in $[a, b]$. The notion of unimodality can be used to reduce the length of the interval containing the optimizing point. For instance, if a unimodal function f is evaluated at any two points x_1, x_2 in $[a, b]$ and if x^* is the minimizing point, then

(i) $x^* \in [x_1, b]$ if $f(x_1) > f(x_2)$;
(ii) $x^* \in [a, x_2]$ if $f(x_1) < f(x_2)$;
(iii) $x^* \in [x_1, x_2]$ or $x^* \in [a, x_2]$ or $x^* \in [x_1, b]$ if $f(x_1) = f(x_2)$.

Here, we have assumed that $x_1 < x_2$. Using this observation, we can prove that if a unimodal function f is evaluated at n points in $x_1 < x_2 < \ldots < x_n$ in $[a, b]$ and

$$f(x_k) = \min_{1 \leqslant i \leqslant n} f(x_i), \tag{8.26}$$

then $x_{k-1} \leqslant x^* \leqslant x_{k+1}$, i.e., the minimizing point is in the interval $[x_{k-1}, x_{k+1}]$.

The foregoing observation has been used in search procedures devised to solve single-variable optimization problems. Search procedures may be broadly categorized into two classes: simultaneous search and sequential search which we now discuss.

8.5.1 SIMULTANEOUS SEARCH PROCEDURES

In the simultaneous search procedure, all the points at which the function is to be evaluated are chosen before the search commences. Any well-defined rule that generates n points in a given interval may be considered a simultaneous search procedure. For example, we may choose n points at random or equally divide the given interval into $n - 1$ subintervals.

If we choose the n points by equally dividing the given interval $[a, b]$ into $n - 1$ subintervals, then by Eq. (8.26), x^* can be determined within an interval of length $[1/(n - 1)](b - a)$. Thus, if we want x^* to be within an interval of length less than ϵ, we can choose the smallest n for which $[1/(n - 1)](b - a) < \epsilon$. This value of n determines our search completely.

8.5.2 SEQUENTIAL SEARCH PROCEDURES

In the sequential search procedure, the search points are generated as the search proceeds.

Dichotomous Search

The dichotomous procedure is based on the fact that if $n = 2$, then x_1, x_2 should be as close as possible to the centre of the given interval. By using unimodality, we can discard at least one of the end intervals. This principle should be repeatedly applied until the prescribed number of function evaluations has been performed. It may be proved that if the function is to be evaluated n times and $x_2 - x_1 = \epsilon$, then the given interval can be reduced by a factor of $2^{-n/2} + (1 - 2^{-n/2})\epsilon$.

A major disadvantage of the dichotomous search is that it does not make full use of the function evaluations.

Fibonacci Search

The Fibonacci procedure is largely dependent on the Fibonacci sequence which is defined as

$$F_0 = F_1 = 1,$$
$$F_k = F_{k-1} + F_{k-2} \quad \text{for } k \geqslant 2.$$

If the function is to be evaluated only n times in $[a, b]$, then let $x_1 = (b - a)(F_{n-1}/F_n) + a$ and locate x_2 symmetrically with respect to x_1 in $[a, b]$, i.e., $x_1 - a = b - x_2$. By using the unimodality of f, we discard either (a, x_1) or (x_2, b), provided $x_1 < x_2$. The interval that remains contains x_1 or x_2. The search is continued by locating the next point symmetrically with respect to the one that already exists in the interval (x_1 or x_2). Thus, by knowing n, the first point can be specified, and this, in turn, specifies the entire Fibonacci search.

It may be proved that x^* lies in an interval of length $(b - a)/F_n$.

Golden Search

The golden search is a simplified version of the Fibonacci search in that it is a good approximation to the Fibonacci search. In the Fibonacci search, the reduction factor at each iteration varies, but it converges to a constant as $n \to \infty$. To be specific,

$$\lim_{n \to \infty} \frac{F_n}{F_{n-1}} = \tau = \frac{1 + \sqrt{5}}{2} = 1.618033989 \ldots .$$

If we retain the principal features of the Fibonacci search, but keep the reduction factor at each iteration as $r = \tau - 1$, we get the golden search.

When the current interval is $[a, b]$, we define

$$x_1 = a + r^2(b - a),$$
$$x_2 = a + r(b - a).$$

It may be observed that x_1 and x_2 are symmetrically placed with respect to $[a, b]$, and the reduction factor is r. Thus, the initial point in the golden search is different from that in the Fibonacci search. The rest of the method, however, is the same for both procedures.

It may be proved that the reduction factor at each iteration is r and therefore, after n iterations, the final interval is of length $r^{n-1}(b - a)$.

Interval Bisection Method

If f is differentiable and unimodal, then

(i) $x^* \in (a, \frac{a + b}{2})$ if $\frac{d}{dx} f(\frac{a + b}{2}) > 0$;

(ii) $x^* \in (\frac{a + b}{2}, b)$ if $\frac{d}{dx} f(\frac{a + b}{2}) < 0$;

(iii) $x^* = \frac{a + b}{2}$ if $\frac{d}{dx} f(\frac{a + b}{2}) = 0$.

These observations imply that a knowledge of the sign of the derivative of a unimodal function is useful in reducing the interval. Thus, after n evaluations of $f'(x)$, the interval $[a, b]$ can be reduced to an interval of length $(0.5)^n(b - a)$.

Interpolation Procedures

The interpolation procedures are based on the following fact. Any continuous function can be approximated to any degree of accuracy by a polynomial. Also, if the function satisfies the unimodality property, then polynomials of a low degree, particularly second-degree polynomials, can be used. This is especially true in the vicinity of the optimizing point where a second-degree polynomial serves as a good approximation. Thus, methods that locate first an interval containing the optimizing point and then the desired point by means of the second-degree or cubic polynomial approximations have been devised. We here describe one such procedure.

Step 1   Initialize, i.e., choose any x_0 (as an estimate of the minimizing point) and step length Δ_0. Set $k = 0$.

Step 2   Compute $x_{k+1} = x_k + \Delta_k$ and $f(x_{k+1})$.

Step 3   If $f(x_{k+1}) \leqslant f(x_k)$, replace Δ_k by $2\Delta_k$ and repeat step 2 with $k \to k + 1$.

Step 4   If $f(x_{k+1}) > f(x_k)$ and $k \geqslant 1$, compute $f(x_{k+1} - \Delta_k/2)$. Of the four equidistant points thus obtained, use three that bracket the minimum. If $f(x_{k+1}) \geqslant f(x_k)$ and $k = 0$, reverse the search direction and compute $f(x_0 - \Delta_0)$. If this fails again, the minimum lies between $x_0 - \Delta_0$ and $x_0 + \Delta_0$. Otherwise, repeat step 2 with $\Delta_1 = -2\Delta_0$ and $k = 1$.

Step 5   When three equidistant points denoted by x_1, x_2, x_3 ($x_1 < x_2 < x_3$) are obtained such that $\Delta = x_2 - x_1 = x_3 - x_2$, a parabola is fitted through $[x_i, f(x_i)]$, $i = 1, 2, 3$. If x_m is the minimizing point of this parabola, then

$$x_m = x_2 - \frac{\Delta}{2}\left[\frac{f(x_3) - f(x_1)}{f(x_1) - 2f(x_2) + f(x_3)}\right] = x_2 + \Delta_m.$$

Step 6   If $|\Delta_m| < \tau$, where τ is the tolerance limit, then x_m is accepted as a minimizing point of $f(X)$. Otherwise, the process is repeated with reduced step length and the point x_2 or x_m, whichever gives the lesser value of $f(x)$.

It may be noted that this procedure searches the whole real line and not merely the interval $[a, b]$. It may also be mentioned that it attempts to locate the minimizing point, whereas the earlier procedures only reduce the interval containing the minimizing point.

Cubic Interpolation Procedure

The problem we shall discuss in Section 8.6 is

$$\min_{X \in R^n} \ f(X).$$

Most of the procedures require the solution of the single-variable problem

$$\min_{\alpha \in R} f(X_0 + \alpha d),$$

where X_0, $d \in R^n$ are given. The function $f(X_0 + \alpha d)$ can be approximated by a cubic polynomial in α and then a solution procedure can be developed. Suppose α_1, α_2 are known such that $\alpha_1 \leqslant \alpha_0 \leqslant \alpha_2$, where α_0 is the minimizing point. By knowing $f(X_0 + \alpha_i d)$ and $\nabla f(X_0 + \alpha_i d)$, $i = 1, 2$, we can approximate $f(X_0 + \alpha d)$ by means of a cubic polynomial. If $\alpha_m = \alpha_1 + \Delta\alpha_m$ is the minimizing point of this cubic polynomial, then a brief calculation shows that

$$\Delta\alpha_m = \Delta\alpha[1 - \frac{\nabla f(\alpha_2)d + W - Z}{\nabla f(\alpha_2)d - \nabla f(\alpha_1)d + 2W}],$$

where

$$\Delta\alpha = \alpha_2 - \alpha_1,$$

$$Z = \frac{3}{\Delta\alpha}[f(\alpha_1) - f(\alpha_2)] + [\nabla f(\alpha_1) + \nabla f(\alpha_2)]d,$$

$$W = [Z^2 - \{\nabla f(\alpha_1)d\}\{\nabla f(\alpha_2)d\}]^{1/2}.$$

Here,

$$f(\alpha_i) = f(X_0 + \alpha_i d),$$

$$\nabla f(\alpha_i) = \nabla f(X_0 + \alpha_i d)$$

A good initial choice of $\Delta\alpha$ is

$$\Delta\alpha = \min\left[1, \frac{-2\{f(\alpha_1) - f(\alpha_0)\}}{\nabla f(\alpha_1)d}\right],$$

where $f(\alpha_0)$ is a lower bound of $f(X_0 + \alpha d)$.

The steps of the cubic interpolation algorithm are now stated.

Step 1   Choose α_1, α_2 such that the minimizing point is contained in $[\alpha_1, \alpha_2]$, and $\Delta\alpha = \alpha_2 - \alpha_1$ is close to the choice just suggested.

Step 2   Compute $f(\alpha_i)$ and $\nabla f(\alpha_i)$, $i = 1, 2$, and thus compute $\alpha_m = \alpha_1 + \Delta\alpha_m$, where α_m is the minimizing point of the cubic polynomial approximating $f(X_0 + \alpha d)$.

Step 3   If $\min\{f(\alpha_1), f(\alpha_2)\} \geqslant f(\alpha_m)$, accept α_m as the minimizing point. Otherwise, compute $\nabla f(\alpha_m)d$ and proceed to step 4.

Step 4   If $\nabla f(\alpha_m)d < 0$, replace α_1 by α_m and return to step 2. Otherwise, replace α_2 by α_m and return to step 2.

In this section, we have described a few of the many search procedures for a single-variable optimization problem. If a search procedure for the problem

$$\text{minimize } f(x), \quad a \leqslant x \leqslant b,$$

evaluates the function f at n points, namely,

$$a \leqslant x_1 < x_2 < \ldots < x_n \leqslant b,$$

and

$$f(x_k) = \min_{1 \leqslant i \leqslant n} f(x_i),$$

then the minimizing point x^* is in (x_{k-1}, x_{k+1}). When several search procedures are applied to the same problem, the interval (x_{k-1}, x_{k+1}) varies with each procedure. But our concern is not as much with the interval as with the length $x_{k+1} - x_{k-1}$.

The following results have been mathematically proved:

(i) The Fibonacci search procedure minimizes the maximum of $x_{k+1} - x_{k-1}$, the maximum being taken over all the quasiconvex functions.

(ii) If a unimodal function is to be minimized, then the maximum interval, which can be reduced to a unit interval containing the minimizing point and using at the most n function evaluations, is F_n.

(iii) If a unimodal function is defined on a discrete set of points over which the minimum is being sought, then the maximum number of points that can be scanned, with at the most n function evaluations, is $F_{n+1} - 1$.

The Fibonacci search is popular because it achieves these results.

8.6 MULTIVARIABLE SEARCH

The penalty function approach, which we shall follow in this text, solves an NLP problem by solving a sequence of unconstrained optimization problems. Accordingly, in this section, we concentrate on solution procedures for the problem

$$\min_{X \in R^n} f(X).$$

We shall now consider two kinds of methods: in one, only the function values are calculated, and in the other, the function values together with the gradients are computed. A number of methods of each kind are available, but we shall confine our discussion to the widely used ones.

8.6.1 SIMPLEX METHOD (Nelder and Mead)

A simplex in R^n is the convex hull of $n + 1$ points $X_0, X_1, \ldots, X_n$ such that $X_i - X_0$, $i = 1$, $2, \ldots, n$, are linearly independent. This is a generalization of the concepts of line segment, triangle, and tetrahedron in one-, two-, and three-dimensional space, respectively. The simplex method starts with a given simplex and then changes it at subsequent iterations in the hope of locating the minimizing point in the final simplex which is of a reasonable size. This procedure is described as follows.

Step 1   Choose any $n + 1$ points $X_0, X_1, \ldots, X_n$ such that $X_i - X_0$, $i = 1, 2, \ldots, n$, are linearly independent.

Step 2   Compute

$$f_i = f(X_i), \qquad i = 1, 2, \ldots, n,$$

$$f_p = \max_{1 \leqslant i \leqslant n} \{f_i\},$$

$$f_m = \min_{1 \leqslant i \leqslant n} \{f_i\},$$

$$\bar{X} = \frac{1}{n} \sum_{\substack{i=0 \\ i \neq p}}^{n} X_i, \qquad \bar{X} \text{ is the centroid of all } X_i \text{ except } X_p.$$

Step 3 X_p is replaced by X_y, X_e, X_c which are obtained by using the geometrical concepts of reflection, expansion, contraction respectively, as now explained:

   (i) $X_y = (1 + \alpha)\bar{X} - \alpha X_p$ (reflection),

provided $f_i > f_y > f_m$ for at least one i other than $i = m$ or $i = p$.

   (ii) $X_e = \gamma X_y + (1 - \gamma)\bar{X}$ (expansion),

provided (a) $f_y < f_e$ and (b) $f_e < f_m$. If (a) holds but not (b), then replace X_p by X_y.

   (iii) $X_c = \beta Y + (1 - \beta)\bar{X}$ (contraction),

where $Y = X_p$ or $Y = X_y$ according as $f_y \geqslant f_p$ or $f_y < f_p$. Here, the choice of α, β, γ is relevant; suggested values are $\alpha = 1$, $\beta = 2.5$, $\gamma = 2$.

Step 4   Return to step 2 with the changed simplex.
    The convergence criterion is

$$\frac{1}{n} \sum_{i=0}^{n} (f_i - \bar{f})^2 \leqslant \epsilon,$$

where ϵ is the tolerance limit, and $\bar{f}$ the mean of all f_i.
    The choice of the initial simplex is important. Let $X_0 = (x_1^0, x_2^0, \ldots, x_n^0)$ be an estimate of the minimizing point. For a given l, let

$$p_n = \frac{\sqrt{n + 1} - 1 + n}{n\sqrt{2}} l, \qquad q_n = \frac{\sqrt{n + 1} - 1}{n\sqrt{2}} l.$$

Then, if

$$X_i = (x_1^i, x_2^i, \ldots, x_n^i), \qquad i = 1, 2, \ldots, n,$$

where

$$x_j^i = x_j^0 + q_n, \qquad j \neq i,$$

$$x_i^i = x_i^0 + P_n,$$

it can be proved that $X_0, \ldots, X_n$ is a simplex with each edge having length l. Depending on our confidence in X_0, we can adjust l to obtain the initial simplex.

8:6.2 PATTERN SEARCH METHOD (Coordinatewise search)

In the pattern search method, the objective function is searched successively along each of the chosen coordinate axes. Finally, if no appreciable improvement is obtained, the ultimate point is claimed to be optimal. The algorithm is described as follows.

Step 1   Choose an estimate X_0^0 of the minimizing point, and a set of n linearly independent vectors $d_1, d_2, \ldots, d_n$ to represent the coordinate axes.

Step 2   Compute

$$\min_{\alpha} f(X_0^0 + \alpha d_1) = f(X_0^1)$$

$$\min_{\alpha} f(X_0^1 + \alpha d_2) = f(X_0^2)$$

$$\vdots$$

$$\min_{\alpha} f(X_0^{n-1} + \alpha d_n) = f(X_0^n)$$

These problems can be solved by using single-variable search methods. In practice, the actual minimum may not be computed. All that we need along the i-th direction is a value of α such that

$$f(X_0^{i-1} + \alpha d_i) \leqslant f(X_0^{i-1}).$$

In fact, for the i-th direction, the step length α_i can be prescribed in advance and we can even compute

$$\min \left[f(X_0^{i-1} + \alpha_i d_i), f(X_0^{i-1} - \alpha_i d_i), f(X_0^{i-1}) \right]$$

and thereby obtain X_0^i starting with X_0^{i-1}.

Step 3   Compute

$$\min_{\alpha} f[X_0^n + \alpha(X_0^n - X_0^0)] = f(X_1^0).$$

Step 4   If $|f(X_0^0) - f(X_1^0)| < \epsilon$, terminate. Otherwise, return to step 2, replacing X_0^0 by X_1^0.

It is obvious that the choice of $d_1, d_2, \ldots, d_n$ is crucial. In fact, for a quadratic function, if $d_1, d_2, \ldots, d_n$ are the principal axes, then X_0^n is the optimal point, irrespective of the choice of X_0^0. Hence, some methods that change the axes whenever the process requires to be repeated have been devised. Of these, we now consider only the gradient methods.

8.6.3 DESCENT METHODS

The basic iterations followed in the descent methods are now listed.

Step 1   Choose an estimate of the minimizing point X_0.

Step 2   Find a direction d_0 such that

$$f(X_0 + \alpha d_0) < f(X_0)$$

for all α satisfying $0 < \alpha \leqslant \bar{\alpha}$ for some $\bar{\alpha} > 0$.

Step 3   Choose α_0 such that $0 < \alpha_0 \leqslant \bar{\alpha}$.

Step 4   If $|f(X_0) - f(X_0 + \alpha_0 d_0)| < \epsilon$, terminate. Otherwise, return to step 2 replacing X_0 by $X_0 + \alpha_0 d_0$.

The choice of d_0 in step 2 and α_0 in step 3 gives rise to various descent methods.

It has been observed that any quasiconvex function in the vicinity of the optimizing point behaves like a quadratic function. Therefore, methods are first devised for minimizing a convex quadratic function, and then extended as approximate methods for an arbitrary quasiconvex function.

The behaviour of a quadratic function can be more easily understood than that of a non-quadratic nonlinear function. For example, if the quadratic function is

$$f(X) = \tfrac{1}{2} X' H X + b X + C,$$

where H is a positive definite matrix, then the minimizing point $\bar{X}$ is given by

$$\bar{X} = X_0 - H^{-1} \nabla f(X_0),$$

where X_0 is any arbitrary point. This shows that if we choose $d_0 = -H^{-1} \nabla f(X_0)$ in step 2 and $\alpha_0 = 1$ in step 3, then a quadratic function can be minimized in one iteration. It should be noted that $H = \|\partial^2 f/\partial x_i\, \partial x_j\|$, i.e., H is the Hessian matrix. For a nonquadratic function, the Hessian matrix is not a constant matrix, a difficulty that can be overcome by considering a certain approximation of the inverse of the Hessian matrix. Each such approximation gives rise to a descent method. Of the several available approximations, we shall discuss here only the one used by Fletcher and Powell.

It should be noted that the following result can be proved : d_0 is a descent direction at X_0 (as explained in step 2) if and only if there exists some positive definite matrix H such that $d_0 = -H \nabla f(X_0)$.

8.6.4 FLETCHER-POWELL METHOD

By using the discussion in Section 8.6.3, we state the Fletcher-Powell method as follows.

Step 1   Start with any X_0 and a positive definite matrix H_0, say $H_0 = I$. Set $k = 0$.

Step 2   Compute $s_k = \nabla f(X_k)$.

Step 3   Compute $d_k = -H_k s_k$.

Step 4   Compute $\min_{\alpha} f(X_k + \theta d_k)$; let this minimum occur at $\theta = \theta_k$.

Step 5   Compute $X_{k+1} = X_k + \theta_k d_k$.

Step 6   If $|f(X_k) - f(X_{k+1})| < \epsilon$, terminate. Otherwise proceed to step 7.

Step 7   Compute

$$s_{k+1} = \nabla f(X_{k+1}),$$

$$\sigma_k = X_{k+1} - X_k,$$

$$Y_k = s_{k+1} - s_k.$$

Step 8   Compute

$$H_{k+1} = H_k - \frac{(H_k Y_k)(H_k Y_k)'}{Y_k' H_k Y_k} + \frac{\sigma_k \sigma_k'}{\sigma_k' Y_k}.$$

Step 9   Return to step 3 replacing k by $k + 1$.

It can be proved that, if this algorithm is applied to a convex quadratic function of n variables, the minimum can be obtained in at the most n iterations. Such an algorithm is said to be *quadratically convergent*.

For computational purposes, the derivatives can be numerically computed by using the finite difference technique and any θ_k for which $f(X_k + \theta_k d_k) < f(X_k)$ can be chosen.

It may be remarked that although the quadratically convergent algorithm uses only first-order partial derivatives, because of the approximation used in step 8, it incorporates most of the features of an algorithm based on second-order partial derivatives. One of the drawbacks of this algorithm is that H_k may become singular for some k. To check this, the determinants can be updated by using the recursion formula

$$|H_k| = |H_{k-1}| \frac{\sigma_{k-1}' Y_k}{Y_{k-1}' G_{k-1} Y_{k-1}}.$$

If the value of $|H_k|$ falls below a certain limit, then H_k can again be replaced by I and the process continued.

In this section, we have given some of the most popular solution procedures used for unconstrained optimization problems. Several other solution procedures are also available. The choice of an appropriate algorithm is therefore very important. We shall now discuss some of the criteria used in the evaluation of algorithms.

(i) The *robustness* of the algorithm, i.e., the ability of the algorithm to solve most of the problems posed. It is obvious that any algorithm can be defeated by a suitably designed (pathological) problem, and hence we should not expect an algorithm to solve *all possible* problems, not even those that are not pathological. This should be borne in mind because an unconstrained optimization problem leading to negative arguments (such as division by zero and discontinuities) is often encountered.

(ii) The *desired degree of precision* in the solution, i.e., in $f(x^*)$ and x^*, where x^* is an optimum solution. This degree of precision depends on the termination criteria used for computation. If the termination of the algorithm depends solely on the fractional changes in $f(x)$ being less than some small number, a flat plateau can cause premature termination. If the termination of the algorithm depends solely on the fractional changes in the elements of x, a steep slope can cause premature termination. If the components of $\nabla f(x)$ are used for the termination of the algorithm, the final point may be only a saddle point. A combination of these three criteria may give the desired degree of precision.

(iii) The *number of functional evaluations* of $f(x)$ for attaining the desired precision in $f(x)$ and x. Since it is difficult to count these evaluations, the *time taken to execute* the algorithm is instead taken as the measure of effectiveness of the algorithm.

Several studies indicate that the Fletcher-Powell algorithm adequately meets most of these criteria.

8.7   CONSTRAINED OPTIMIZATION (Penalty Function Approach)

The problem we consider in this section is the general NLP problem:

$$\text{Minimize } f(X) \tag{8.27}$$

subject to

$$g_i(X) \leqslant 0 \qquad (i = 1, 2, \ldots, m). \tag{8.28}$$

It should be noted that the non-negativity restrictions, if any, on the variables have also been included in constraint (8.28). Let

$$S = \{X \mid g_i(X) \leqslant 0, \, i = 1, 2, \ldots, m\},$$

i.e., S is the feasible region. It is obvious that this inequality-constrained optimization problem is equivalent to the unconstrained optimization problem

$$\min_{X \in R^n} \{f(X) + P(X)\}, \tag{8.29}$$

where

$$P(X) = \begin{cases} 0 & \text{if } X \in S \\ \infty & \text{if } X \notin S \end{cases}$$

The function $P(X)$ can be considered a penalty function which imposes a very heavy penalty for leaving the feasible region. In a structural design problem, leaving the feasible region is equivalent to violating a constraint. In view of this character of $P(X)$, it is obvious that the methods of unconstrained optimization, discussed in Section 8.6, cannot be immediately applied to the problem defined by (8.29). To circumvent this difficulty, a sequence of increasingly complex problems, which, in the limit, converge to (8.29), can be obtained. This is done by defining a sequence of functions $P_k(X)$ which converges to $P(X)$ as $k \to \infty$. Also, when $P(X)$ in (8.29) is replaced by $P_k(X)$, the problem can be solved by the methods given in Section 8.6. This means that the inequality-constrained optimization problem can be solved by means of a sequence of unconstrained optimization problems.

Several alternative sequences of $P_k(X) \to P(X)$ are available, but we restrict our choice to the sequence that is particularly suited to structural design problems. Such a sequence is given by

$$P_k(X) = -r_k \sum_{i=1}^{m} \frac{1}{g_i(X)},$$

where $r_k > r_{k+1} > 0$. Thus, the sequence of unconstrained optimization problems that we need to consider is given by

$$\min_{X \in R^n} \left\{ f(X) - r_k \sum_{i=1}^{m} \frac{1}{g_i(X)} \right\}, \tag{8.30}$$

where $\{r_k\}$ is a decreasing sequence of positive real numbers tending to zero.

For the problem defined by (8.27), (8.28), we assume that

(i) $S^0 = \{X \mid g_i(X) < 0, \, i = 1, 2, \ldots, m\} \neq \phi$, i.e., there is at least one feasible solution where none of the constraints is active;

(ii) $f(X)$ has a finite minimum over the feasible region S;

(iii) the functions f and g_i are twice continuously differentiable;

(iv) the functions f and g_i are convex and at least one of them is strictly convex.

When these conditions are imposed, (8.27), (8.28) is a well behaved problem where a smooth convex function is to be minimized over a convex set with smooth boundaries. For such a problem, the following *result* is proved.

If $X(r_k)$ is a solution of problem (8.30) and X^* is a solution of (8.27), (8.28), then

$$\lim_{r_k \to 0} X(r_k) = X^*.$$

In practice, however, it is extremely difficult to verify whether conditions (i)-(iv), just listed, are satisfied. Even when these conditions are not satisfied, experience has shown that SUMT still gives solutions that may turn out to be only local minima and not global minima.

The question that obviously arises is: Why cannot we minimize $P_k(X)$ for a very small r_k and claim the solution so obtained to be that of the original problem? The answer lies in the fact that it is difficult to minimize $P_k(X)$ for a very small r_k. To overcome this difficulty, $P_k(X)$ is initially minimized for some 'reasonable' value of r_k and then the solution thus obtained is used as the starting solution when r_k is further reduced. Thus, ultimately $P_k(X)$ has to be minimized for a small value of r_k; the difficulty involved during such minimization is overcome sequentially in the earlier problems where r_k is not small.

The foregoing discussion shows that the *initial choice of* r_0 and reduction factors at the subsequent stages are very important. Let us assume that a point X_0, where $g_i(X_0) < 0$, $i = 1, 2, \ldots, m$, exists. If X_0 is not 'near' any boundary, then the initial $r = r_0$ can be computed from the equation

$$f(X_0) = -r_0 \sum_{i=1}^{m} \frac{1}{g_i(X_0)}. \tag{8.31}$$

If $g_i(X_0)$ is nearly zero for at least one i, then r_0 is chosen as a very large number. Another choice of r_0 is given by

$$r_0 = \frac{\nabla f(X_0)' \nabla p(X_0)}{[\nabla p(X_0)]^2}, \tag{8.32}$$

where

$$p(X) = \sum_{i=1}^{m} \frac{1}{g_i(X)}.$$

The reduction factor c, defined by $r_k = c r_{k+1}$, is usually chosen between 10 and 50. Though this factor can be large initially, it has subsequently to be reduced when $P_k(X)$ becomes problematic.

If we consider the unconstrained problems corresponding to r_k and r_{k+1}, i.e., at two consecutive iterations, then

$$f(X_{k+1}) - r_{k+1} P(X_{k+1}) < f(X_k) - r_{k+1} P(X_k), \tag{8.33}$$

$$f(X_k) - r_k P(X_k) < f(X_{k+1}) - r_k P(X_{k+1}), \tag{8.34}$$

where

$$P(X) = \sum_{i=1}^{m} \frac{1}{g_i(X)}.$$

On a rearrangement of the terms, inequality (8.33) gives

$$f(X_{k+1}) - f(X_k) < r_{k+1}[P(X_{k+1}) - P(X_k)]. \tag{8.35}$$

Similarly, from inequality (8.34), we get

$$r_k[P(X_{k+1}) - P(X_k)] < f(X_{k+1}) - f(X_k). \tag{8.36}$$

Using inequality (8.35) in inequality (8.36), we obtain

$$r_k[P(X_{k+1}) - P(X_k)] < r_{k+1}[P(X_{k+1}) - P(X_k)]$$

which gives

$$P(X_{k+1}) < P(X_k) \tag{8.37}$$

since $r_k > r_{k+1}$. Using inequality (8.37) in inequality (8.35), we get

$$f(X_{k+1}) < f(X_k). \tag{8.38}$$

Inequality (8.37) ensures that the unconstrained objective function tends to the original objective function. Inequality (8.38) makes sure that, at each subsequent iteration, a better feasible point is obtained.

By using the duality theory of NLP, another result proved is

$$f(X_k) + r_k \sum_{i=1}^{m} \frac{1}{g_i(X_i)} \leqslant f(X^*) \leqslant f(X_k), \tag{8.39}$$

where X^* is the solution of (8.27), (8.28). Inequality (8.39) gives the error bounds which can be used as termination criteria. Hence, if

$$\left| r_k \sum_{i=1}^{m} \frac{1}{g_i(X_k)} \right| < \epsilon,$$

where ϵ is the tolerance limit, then we can terminate the computations.

Another termination criterion can be obtained from the approximation of the function $f[X(r)]$ by using the data accumulated in two or more successive iterations. Experience indicates that $f[X(r)]$ can be approximated by

$$f[X(r)] \approx a + b\sqrt{r}. \tag{8.40}$$

If we apply this approximation when $r = r_k$ and $r = r_{k+1} = r_k/c$, the two equations we obtain give

$$f[X(0)] \approx a = \frac{1}{\sqrt{c}-1} \{ \sqrt{c} f[X(r_{k+1})] - f[X(r_k)] \}. \tag{8.41}$$

Thus, if at any iteration $f[X(r_k)]$ is close to $f[X(0)]$ as computed by Eq. (8.41), the algorithm may be terminated.

8.7.1 USE OF EXTRAPOLATION

It may be proved that $X(r)$ which minimizes the unconstrained function

$$f(X) - r \sum_{i=1}^{m} \frac{1}{g_i(X)}$$

is a continuous function of r. Thus, by using the value of $X(r)$ at $r = r_k$ and $r = r_{k+1} = r_k/c$ an extrapolated vector $X_E(r)$ can be obtained. This vector, if feasible, may be used as a starting point for the next unconstrained minimization. Experience shows that this tactic substantially accelerates the computational procedure. To be precise, assume that

$$X_E(r) = a + b\sqrt{r}, \tag{8.42}$$

where a and b are now vectors. Using the computed value of $X(r_k)$ and $X(r_{k+1})$, where $cr_{k+1} = r_k$, we get

$$a = \frac{1}{\sqrt{c} - 1}[\sqrt{c}X(r_{k+1}) - X(r_k)], \tag{8.43}$$

$$b = \frac{X(r_k) - a}{\sqrt{r_k}}. \tag{8.44}$$

By substituting Eqs. (8.43) and (8.44) in (8.42), $X_E(r_{k+2})$ can be computed. If $X_E(r_{k+2})$ is feasible, it provides a good starting point.

BIBLIOGRAPHY

Box, M. J., Davies, D., and Swann, W. H., Nonlinear Optimization Techniques, ICI Monograph No. 5, Oliver and Boyd, Edinburgh, 1969.

Broyden, C. G., A class of methods for solving nonlinear equations, *Math. Comp.*, **19**, 92, 577–84, 1965.

Fletcher, R., Function minimization without evaluating derivatives—a review, *Computer J.*, **8**, 1, 33–41, 1965.

Fletcher, R., A new approach to variable metric algorithms, *Computer J.*, **13**, 4, 317–22, 1970.

Fletcher, R. and Powell, M. J. D., A rapidly convergent descent method for minimization, *Computer J.*, **6**, 2, 163–68, 1963.

Fletcher, R. and Reeves, C. M., Function minimization by conjugate gradients, *Computer J.*, **7**, 2, 149–54, 1964.

Goldstein, A. A. and Price, J. F., An effective algorithm for minimization, *Numer. Maths.*, **10**, 184–89, 1967.

Hooke, R. and Jeeves, T. A., Direct search solution of numerical and statistical problems, *J. ACM*, **8**, 212–21, 1961.

Nedler, J. A. and Mead, R., A simplex method for function minimization, *Computer J.*, **7**, 4, 308–13, 1965.

Pearson, J. D., Variable metric methods of minimization, *Computer J.*, **12**, 2, 171–78, 1969.

Powell, M. J. D., An efficient method for finding the minimum of a function of several variables without calculating derivatives, *Computer J.*, **7**, 2, 155–62, 1964.

Rosenbrock, H. H., An automatic method of finding the greatest or least value of a function, *Computer J.*, **3**, 3, 175–84, 1960.

Stewart, III, G. W., A modification of Davidon's minimization method to accept difference approximations to derivatives, *J. ACM*, **14**, 1, 72–83, 1967.

Zangwill, W. I., Minimizing a function without calculating derivatives, *Computer J.*, **10**, 3, 293–96, 1967.

9
Geometric Programming

9.1 INTRODUCTION

Geometric programming (GP) is a technique applicable to a particular class of nonlinear programming (NLP) problems where both the objective function and the constraint functions are *posynomials* (see Section 9.2). This technique was developed in the 1960s and was initially applied to problems in chemical and electrical engineering. Recently, it has been effectively employed also in structural engineering.

The GP technique is based essentially on a classical inequality, called the *arithmetic-geometric-mean inequality*. When this inequality is used, the geometric programme, together with the objective function having nonlinear constraints, is reduced to an equivalent nonlinear programme where the constraints are linear. Attempts are currently being made to solve this particular type of NLP problem.

A novel feature of the GP technique is that it aims at computing first the minimum value of the objective function and then the optimum values of the variables. What is more, it gives the optimum values of various terms comprising the objective function even before the optimum values of the variables are worked out. In design problems, these terms refer to the costs of various components of the structure. Thus, even without knowing the optimum values of the design parameters, the optimum values of the components of the structure can be computed. The knowledge of these values is useful in, for example, feasibility studies and tendering. If the GP problem cannot be easily solved, bounds for the optimum objective function value can always be obtained by employing the GP technique.

In Section 9.2, we describe and illustrate the basic GP technique for unconstrained geometric programmes, and in Section 9.3, we consider the technique for constrained geometric programmes.

9.2 UNCONSTRAINED GEOMETRIC PROGRAMMES

As already noted, a geometric programme is a nonlinear programme in which the problem functions are generalized polynomials with positive coefficients and variables and arbitrary real exponents. Such generalized polynomials are called *posynomials*. A general posynomial is therefore of the form

$$f(X) = \sum_{j=1}^{m} C_j \left(\prod_{i=1}^{n} X_i^{a_{ji}} \right), \tag{9.1}$$

where, for all i and j, $C_j > 0$, $X_i \geqslant 0$. For example,

$$f(X) = 10x_1^3 x_2^{-1} x_3^{3.5} + 7.6x_1^{-4.3} x_2^{3.7}$$

is a posynomial in three variables.

We now consider the GP technique for minimizing a function of the type given by Eq.

(9.1). This technique is based on the following weighted arithmetic-geometric-mean inequality. If $u_1, u_2, \ldots, u_m$ are any m positive real numbers, then

$$u_1 + u_2 + \ldots + u_m \geqslant \left(\frac{u_1}{V_1}\right)^{V_1}\left(\frac{u_2}{V_2}\right)^{V_2} \ldots \left(\frac{u_m}{V_m}\right)^{V_m}, \tag{9.2}$$

where $V_1, V_2, \ldots, V_m$ are any positive real numbers satisfying

$$V_1 + V_2 + \ldots + V_m = 1. \tag{9.3}$$

If we put

$$u_j = u_j(X) = C_j \prod_{i=1}^{n} X_i^{a_{ji}} \tag{9.4}$$

in (9.1) so that

$$f(X) = u_1(X) + u_2(X) + \ldots + u_n(X) \tag{9.5}$$

and use inequality (9.2), we get

$$f(X) \geqslant g(X, V), \tag{9.6}$$

where

$$g(X, V) = \left(\frac{C_1}{V_1}\right)^{V_1}\left(\frac{C_2}{V_2}\right)^{V_2} \ldots \left(\frac{C_m}{V_m}\right)^{V_m} X_1^{D_1} X_2^{D_2} \ldots X_n^{D_n}, \tag{9.7}$$

$$D_i = \sum_{j=1}^{m} a_{ij} V_j \qquad (i = 1, 2, \ldots, n), \tag{9.8}$$

$$\sum_{j=1}^{m} V_j = 1 \qquad (V_j \geqslant 0). \tag{9.9}$$

If we choose the weights V_j satisfying Eq. (9.9) and

$$D_i = \sum_{j=1}^{m} a_{ij} V_j = 0 \qquad (i = 1, 2, \ldots, n), \tag{9.10}$$

then $g(X, V)$ becomes independent of X, and inequality (9.6) gives

$$f(X) \geqslant \bar{g}(V), \tag{9.11}$$

where $\bar{g}(V) = g(X, V)$ for any choice of V satisfying Eqs. (9.9) and (9.10).

We can now prove that if $f(X)$ has a minimum value over $X \geqslant 0$, then $\bar{g}(V)$ has a maximum value when V satisfies (9.9) and (9.10). Also, the maximum value of $\bar{g}(V)$ equals the minimum value of $f(X)$. Thus, if $\bar{g}(V)$ can be maximized subject to the linear equality constraints (9.9) and (9.10), then the minimum value of $f(X)$ is obtained without knowing the optimum values of X_i. For example, if these two equations have a unique solution, then we know the maximum value of $\bar{g}(V)$.

Difficulties, however, arise if the equations do not have a unique solution. In such a situation, we may maximize $\bar{g}(V)$ subject to the equations. Methods are currently being proposed to solve such problems. However, to find a feasible solution of (9.9) and (9.10) is not difficult. Any such feasible solution can be used in inequality (9.11) to obtain the lower bounds on $f(X)$.

To compute the optimum values of the X_i variables, the following result has been proved. If X^0 minimizes $f(X)$ and V^0 maximizes $\bar{g}(V)$ subject to (9.9) and (9.10), then

$$u_j(X^0) = \bar{g}(V^0)V_j^0 \qquad (j = 1, 2, \ldots, m). \tag{9.12}$$

Equation (9.12) can be solved to obtain X^0. For instance, if we take the logarithm of both sides of (9.12), we get linear equations in $\ln X_j^0$. By knowing $\ln X_j^0$, we can compute X_j^0.

It should be observed that (9.9), (9.10) is a system of $n + 1$ linear equations in m positive unknowns. Here, n is the *number of variables* in $f(X)$ and m the *number of terms* in $f(X)$. Therefore, if $m = n + 1$, we expect a unique solution to exist. If $m > n + 1$, then (9.9), (9.10) can be solved for $n + 1$ variables as a linear combination of the remaining $m - (n + 1)$ variables. Thus, $\bar{g}(V)$ can be considered a function of the $m - (n + 1)$ variables and then it can be maximized, keeping in view the fact that we are looking only for positive solutions. In practice, $m - (n + 1)$ is not very large. This accounts for the success of GP. It is therefore worthwhile to know whether maximizing $\bar{g}(V)$ would be computationally useful, and if so when. However, since the studies on GP have not yet developed sufficiently in this direction, we shall not discuss this aspect.

The following example illustrates the GP technique for the unconstrained minimization of a posynomial.

Example 9.1

$$\text{Minimize } f(X) = C_1 x_1^{-2} x_2^{-1} + C_2 x_1 x_2 + C_3 x_1^2. \tag{9.13}$$

The maximization problem corresponding to Eq. (9.13) as obtained from Eqs. (9.7), (9.9), and (9.10) is:

$$\text{Maximize } \bar{g}(V) = \left(\frac{C_1}{V_1}\right)^{V_1}\left(\frac{C_2}{V_2}\right)^{V_2}\left(\frac{C_3}{V_3}\right)^{V_3} \tag{9.14}$$

subject to

$$\begin{aligned}
-2V_1 + V_2 + 2V_3 &= 0, \\
-V_1 + V_2 &= 0, \\
V_1 + V_2 + V_3 &= 1, \\
V_i &\geq 0.
\end{aligned} \tag{9.15}$$

Constraints (9.15) have a unique solution given by

$$(V_1, V_2, V_3) = (\tfrac{2}{5}, \tfrac{2}{5}, \tfrac{1}{5}).$$

Thus, the maximum value of $\bar{g}(V)$ given by Eq. (9.14) subject to (9.15) is

$$\max \bar{g}(V) = \left(\frac{5C_1}{2}\right)^{2/5}\left(\frac{5C_2}{2}\right)^{2/5}(5C_3)^{1/5}. \tag{9.16}$$

A problem where (9.9) and (9.10) possess a unique solution is called a problem with *zero degree of difficulty*. However, if we add one more term to $f(X)$, as given by (9.13), the problem has a *nonzero degree of difficulty*. The example we now give illustrates this situation:

$$\text{Minimize } f(X) = C_1 x_1^{-2} x_2^{-1} + C_2 x_1 x_2 + C_3 x_1^2 + C_4 x_1^2 x_2. \tag{9.17}$$

It should be noted that Eq. (9.17) has resulted by adding $C_4 x_1^2 x_2$ to $f(X)$ in (9.13). Now the maximization problem becomes:

$$\text{Maximize } \bar{g}(V) = \left(\frac{C_1}{V_1}\right)^{V_1}\left(\frac{C_2}{V_2}\right)^{1/2}\left(\frac{C_3}{V_3}\right)^{1/3}\left(\frac{C_4}{V_4}\right)^{1/4} \tag{9.18}$$

subject to

$$-2V_1 + V_2 + 2V_3 + 2V_4 = 0,$$
$$-V_1 + V_2 + V_4 = 0,$$
$$V_1 + V_2 + V_3 + V_4 = 1,$$
$$V_i \geqslant 0.$$

(9.19)

Here, we have three linear equations in four unknowns. We can therefore solve for any three of these variables in terms of the fourth variable. One such solution is given by

$$V_2 = 2 - 4V_1,$$
$$V_3 = 1 - 2V_1,$$
$$V_4 = -2 + 5V_1.$$

(9.20)

The positivity of the variables can be ensured by imposing the restrictions $\frac{2}{5} \leqslant V_1 \leqslant \frac{1}{2}$. Thus, problem (9.18), (9.19) becomes:

$$\text{Maximize } (\frac{C_1}{V_1})^{V_1}(\frac{C_2}{2 - 4V_1})^{2-4V_1}(\frac{C_3}{1 - 2V_1})^{1-2V_1}(\frac{C_4}{5V_1 - 2})^{5V_1-2}$$

subject to

$$\frac{2}{5} \leqslant V_1 \leqslant \frac{1}{2}.$$

This problem can be solved by maximizing the logarithm of the objective function. The solution gives the minimum value of $f(X)$. To obtain the optimum values of the X variables, (9.12) can be used.

It may be observed from the foregoing examples that, even without knowing the exact values of C_i, substantial information on the original problem can be obtained. For this, all that we need to know is the powers of the X variables in the various terms of the objective function to be minimized. Thus, in structural optimization if we know the dimensions of the various terms in the objective function, the maximizing problem can be formulated. Any information that we get from this maximizing problem can be used for the original problem, which we have not yet completely formulated. This aspect of GP needs to be exploited.

9.3   CONSTRAINED GEOMETRIC PROGRAMMES

In Section 9.2, we have explained the basic principle of the GP technique in the context of unconstrained optimization of a posynomial. This technique has been generalized to solve constrained geometric programmes, called *prototype geometric programmes*. It has recently been further so generalized that it can be applied to problems where the C_i's need not be positive. Such problems are called *signomial programming problems*.

As in Section 9.2, a problem of the type (9.7)-(9.10) can be associated with a signomial programme. We now briefly discuss this.

A generalized posynomial, called signomial, is a function of the type

$$f(X) = \sum_{j=1}^{m} \sigma_j C_j (\prod_{i=1}^{n} X_i^{a_{ji}}),$$

(9.21)

where each $\sigma_j = \pm 1$. Thus, a signomial differs from a posynomial in that the various terms of the former need not have positive coefficients. The problem that we are considering is:

$$\text{Minimize } f_0(X)$$

(9.22)

subject to

$$f_p(X) \leqslant \sigma_p \qquad (p = 1, 2, \ldots, m),$$
$$X \geqslant 0,$$

(9.23)

where each $f_p(X)$ is a generalized posynomial and $\sigma_p = \pm 1$. Let

$$f_p(X) = \sum_{j=1}^{m_p} \sigma_{pj} C_{pj} \prod_{i=1}^{n} X_i^{a_{pji}}.$$

(9.24)

For Eq. (9.24), the maximizing problem corresponding to (9.7)–(9.10) is:

$$\text{Maximize } \bar{g}(V) = \sigma[\prod_{p=0}^{m} \prod_{j=1}^{m_p} (\frac{C_{pj} V_{p0}}{V_{pj}})^{\sigma_{pj} \cdot V_{pj}}]^{\sigma}$$

(9.25)

subject to

$$\sum_{j=0}^{m_0} \sigma_{0j} V_{0j} = \sigma (\equiv \pm 1),$$

(9.26)

$$\sum_{p=0}^{m} \sum_{j=1}^{m_p} \sigma_{pj} a_{pji} V_{pj} = 0 \qquad (i = 1, 2, \ldots, n),$$

(9.27)

$$V_{p0} = \sigma_p \sum_{j=1}^{m_p} \sigma_{pj} V_{pj} \geqslant 0 \qquad (p = 1, 2, \ldots, m),$$

(9.28)

$$V_{pj} \geqslant 0 \qquad (p = 0, 1, \ldots, m, j = 1, 2, \ldots, m_p),$$

(9.29)

$$V_{00} = 1,$$

(9.30)

$$\lim_{V_{pj} \to 0} (\frac{C_{pj} V_{p0}}{V_{pj}})^{\sigma_{pj} \cdot V_{pj}} = 0.$$

(9.31)

For problems (9.22)–(9.23) and (9.25)–(9.31), the following result has been proved.

If X^0 is any local minimum of $f_0(X)$ subject to constraints (9.23), then there exist V_{pj}^0 satisfying restrictions (9.26)–(9.31) such that

$$f_0(X^0) = \bar{g}(V^0).$$

(9.32)

Furthermore, $\bar{g}(V)$ is stationary at V^0.

Result (9.32) applies in particular to the global minimum of (9.22), (9.23). Once the V_{pj}^0 satisfying (9.32) corresponding to the global minimum X^0 are known, X^0 can be computed from the equations

$$C_{0j} \prod_{i=1}^{n} X_i^{a_{0ji}} = V_{0j} \sigma \bar{g}(V^0), \qquad j = 1, 2, \ldots, m_0,$$

(9.33)

$$C_{pj} \prod_{i=1}^{n} X_i^{a_{pji}} = \frac{V_{pj}}{V_{p0}}, \qquad j = 1, 2, \ldots, m_p, p = 1, 2, \ldots, m.$$

(9.34)

The following example is intended to explain the use of (9.32).

Example 9.2   Find the cross-sectional dimensions of a structural beam of rectangular cross-section, with breadth x_1 and depth x_2, so as to minimize the cost of the beam whose length is fixed. The cost of the beam is made up of two elements, one directly proportional to the area of the cross-section with a unit cost of Rs C_1 per square inch and the other directly proportional to the depth of the beam with a unit cost of Rs C_2 per inch. Investigations* have demonstrated that any cross-section with a second moment of area $x_1 x_2^3/12$, of not less than r inches, satisfies the structural behaviour requirements.

The cost of the beam is

$$f_0(x_1, x_2) = C_1 x_1 x_2 + C_2 x_2. \tag{9.35}$$

The second moment of area constraint requires

$$\frac{x_1 x_2^3}{12} \geqslant r, \tag{9.36}$$

that is,

$$f_1(x_1, x_2) = 12r x_1^{-1} x_2^{3} \leqslant 1, $$
$$x_1 \geqslant 0, \, x_2 \geqslant 0. \tag{9.37}$$

The problem is to minimize $f_0(x_1, x_2)$ subject to constraints (9.37). The corresponding maximization problem, using (9.25)–(9.29), is

$$\text{maximize } \overline{g}(V) = \left(\frac{C_1}{V_{01}}\right)^{V_{01}} \left(\frac{C_2}{V_{02}}\right)^{V_{02}} (12r)^{V_{11}} \tag{9.38}$$

subject to

$$V_{01} + V_{02} = 1,$$
$$V_{01} - V_{11} = 0,$$
$$V_{01} + V_{02} - 3V_{11} = 0,$$
$$V_{ij} \geqslant 0. \tag{9.39}$$

The unique solution of constraints (9.39) is

$$V^0 = (V_{01}^0, \, V_{02}^0, \, V_{11}^0) = (\tfrac{1}{3}, \, \tfrac{2}{3}, \, \tfrac{1}{3}).$$

Thus, the maximum value of $\overline{g}(V)$ is $\overline{g}(V^0)$. Using (9.32), we see that the minimum value of $f_0(x_1, x_2)$ subject to (9.37) is $\overline{g}(V^0)$. The optimal values of x_1, x_2 can be computed by solving the equation, obtained from (9.33) and (9.34),

$$\begin{bmatrix} 1 & 1 \\ 0 & 1 \\ -1 & -3 \end{bmatrix} \begin{bmatrix} \ln x_1 \\ \ln x_2 \end{bmatrix} = \begin{bmatrix} \ln\left[\tfrac{1}{3}\overline{g}(V^0)\right] \\ \ln\left[\tfrac{2}{3}\overline{g}(V^0)\right] \\ \ln\left[1/(12r)\right] \end{bmatrix}.$$

BIBLIOGRAPHY

Beightler, C. S. and Phillips, D. T., Applied Geometric Programming, Wiley, New York, 1976.
Duffin, R. J., Peterson, E. L., and Zener, C., Geometric Programming: Theory and applications, Wiley, New York, 1967.

*See J. Bracken and G. P. McCormick, Selected Applications of Nonlinear Programming, Wiley, New York, 1968

Morris, A. J., Structural optimization by geometric programming, *Int. J. Solids and Structures*, **8**, 7, 847–64, 1972.

Templeman, A. B., Geometric programming with examples of the optimum design of floor and roof systems, Proc. Int. Symp. Computer-Aided Struct. Des., Warwick University, Peter Peregrinus Ltd., Hartfordshire, 1972.

Templeman, A. B., Wilson, A. J., and Winterbottom, S. K., SIGNPOT—A computer code for the solution of signomial geometric programming problems, Civil Engineering Department Report, Liverpool University, 1972.

10
Dynamic Programming

10.1 INTRODUCTION

In Chapters 7–9, we have given some solution procedures for certain classes of optimization problems for which the now classical techniques of differential calculus and inequalities cannot be used. During the last three decades, another broad category of optimization problems (arising in industry, economics, engineering, mathematics, and social sciences) for which even the techniques of NLP and of calculus of variations are not very helpful has been recognized. This category of problems is referred to as *multistage decision processes*. In this chapter, we shall concentrate on this class of problems and illustrate the basic principles for solving them. The solution procedure for these problems, which was developed by Richard Bellman, is called *dynamic programming*.

10.2 MULTISTAGE DECISION PROCESSES

Consider a system described, at time t, by a finite number of variables $x_1, x_2, \ldots, x_n$. The values of these variables, called *state variables*, at time t, are denoted by $x_1(t), x_2(t), \ldots, x_n(t)$. During the time under consideration, that is, $a \leqslant t \leqslant b$, the system is observed at prescribed times $t_1, t_2, \ldots, t_N$, which, for the sake of simplicity, are denoted by $1, 2, \ldots, N$. Each such time is called a *stage* of the system, t_i being the i-th stage. At each stage i, we have to take a *decision* $d(i)$ from amongst a number of possible choices. We assume here that a single decision is taken at each stage. If we take decision $d(i)$ at stage i, when the system is described by $x_1(i), x_2(i), \ldots, x_n(i)$, the state of the system is so changed that, at stage $i + 1$, the system state is described by

$$x_j(i + 1) = G_j[x_1(i), x_2(i), \ldots, x_n(i), d(i), i], \qquad j = 1, 2, \ldots, n, \, i = 1, 2, \ldots, N - 1,$$
(10.1)

that is, we assume that the *system state* at stage $i + 1$ depends on only (a) the system state at stage i, (b) the decision taken at stage i, and (c) the stage i itself.

With each stage i of the system, there is associated a *stage return function* of the form

$$R(i) = R[x_1(i), x_2(i), \ldots, x_n(i), d(i), i].$$
(10.2)

Further, at each stage i, restrictions that need to be satisfied by the state variables $x_1(i), x_2(i), \ldots, x_n(i)$, and the decision variable $d(i)$ may be imposed. Let us assume that these restrictions are of the form

$$P[x_1(i), x_2(i), \ldots, x_n(i), d(i), i] \lessgtr 0.$$
(10.3)

In addition, we may limit the state and/or decision variables to integer values.

With each *sequence of N decisions* $d(1), \ldots, d(N)$, a system return function I is associated by a relation of the type

I = system return function

$$= I[R(1), R(2), \ldots, R(N)]. \tag{10.4}$$

Usually, I is of the form

$$I = \sum_{j=1}^{N} R(j)$$

or

$$I = \prod_{j=1}^{N} R(j)$$

or

$$I = \max [R(1), R(2), \ldots, R(N)]$$

or

$$I = \min [R(1), R(2), \ldots, R(N)]$$

or

$$I = \text{HCF} [R(1), R(2), \ldots, R(N)]$$

or

$$I = \text{LCM} [R(1), R(2), \ldots, R(N)]$$

and so on.

The problem may be stated as follows. Given the initial state of the system, we are required to choose a feasible sequence of decisions which maximizes or minimizes the system return function.

An optimization problem of the foregoing type is called a *deterministic, discrete-time, finite, multistage* decision process. It is also termed *stationary* if each of the problem functions G_j, R, and P is independent of i.

The essential purpose of the assumptions just given is to enable us to obtain the solution of an $(r + 1)$-stage problem—when the solution of any r-stage problem is known—by solving a single-variable optimization problem. An N-stage problem may thus be decoupled and the solutions of the decoupled problems synthesized to get the desired solution.

10.3   DYNAMIC PROGRAMMING TECHNIQUE

When the problem functions G_j, R, P, and I satisfy the assumptions given in Section 10.2, a solution procedure, called *dynamic programming*, becomes applicable. The dynamic programming technique depends on the *principle of optimality*, which states, according to Bellman, that whatever the initial state of the system and the initial decisions, the subsequent decisions must constitute an optimal policy with regard to the state resulting from the first decision.

Suppose our aim is to maximize the system return function. Evidently, once the problem functions G_j, R, P, and I are fixed, the maximum value of I will depend on N (the total number of stages to be considered) and $x_1(1), x_2(1), \ldots, x_n(1)$ (the initial state of the system). For the sake of convenience, we now introduce the vector notation

$$X(i) = [x_1(i), x_2(i), \ldots, x_n(i)],$$

$$G[X(i), d(i), i] = [G_1\{X(i), d(i), i\}, \ldots, G_n\{X(i), d(i), i\}].$$

Let us assume that we are considering a stationary process. Since the maximum value of I depends on N and $X(1)$, we define

$$f_N[X(1)] = \text{maximum value of } I \text{ over } N \text{ stages}$$
$$\text{when the initial state is } X(1).$$

If the initial decision $d(1)$ is taken arbitrarily, then, corresponding to it, we have

$$X(2) = G[X(1), d(1)];$$

the optimal return over the *remaining* $N - 1$ stages given by

$$f_{N-1}[G\{X(1), d(1)\}];$$

and the return from the first stage given by

$$R[X(1), d(1)].$$

Thus, when the initial decision $d(1)$ is arbitrary, and the remaining $N - 1$ decisions are optimal with respect to the state obtained due to the first arbitrary decision, the total return is

$$R[X(1), d(1)] + f_{N-1}[G\{X(1), d(1)\}].$$

Applying the principle of optimality, we get

$$f_N[X(1)] = \max_{d(1)} \left[R\{X(1), d(1)\} + f_{N-1}[G\{X(1), d(1)\}] \right]. \tag{10.5}$$

Also,

$$f_1[X(1)] = \max_{d(1)} R[X(1), d(1)]. \tag{10.6}$$

Equations (10.5) and (10.6) can be used recursively to solve the problem for any given N and $X(1)$. We may also observe that, at every iteration, we have to solve a maximization problem in one variable only. To obtain a solution of the given problem, the solutions of these single-variable problems are synthesized by using (10.5).

It is evident that the application of the technique of dynamic programming to problems requiring a multistage decision process is comparable to (a) the diagonalization of matrices in a system of linear equations; (b) the separation of variables in partial differential equations; and (c) the use of orthonormal functions in eigenvalue problems. In all these procedures, the attempt is not to meet n difficulties simultaneously but rather to tackle one difficulty at a time.

In this section, we have shown how the dynamic programming technique may be applied to the simplest type of multistage optimization problem. The technique can be successfully used also in intricate problems. To do this, each problem has first to be cast in the format just stated.

We now illustrate the dynamic programming technique by applying it to a machine replacement problem, where we have to make the most suitable compromise between the cost of buying new machinery and the reduced profitability in running old machines.

Example 10.1   Consider a machine about which we have the following information:

$p(i)$ = productivity value of running an i-year-old machine for one year,

$m(i)$ = maintenance cost of running an i-year-old machine for one year,

$s(i)$ = salvage value of an i-year-old machine,

p = purchase price of the new machine.

At the commencement of each year, we have to decide whether to

    (i) keep the machine : K

or

    (ii) replace the machine : R.

Given the initial age of the machine, we are required to determine a sequence of decisions over N periods so that the overall profit is maximized. Assume that the machine is of no use after N periods.

In terms of our notation, we may state that

    (i) the machine constitutes our *system*;

    (ii) the age of the machine $i = 0, 1, 2 \ldots$ denotes the *state of the system* $x(i)$;

    (iii) the state transformation function G is defined by

$$G[i, K] = i + 1,$$

$$G[i, R] = 1;$$

    (iv) the stage return function R, when the machine is i-period-old, is given by

$$R[i, K] = p(i) - m(i),$$

$$R[i, R] = p(0) - m(0) + s(i) - p;$$

    (v) the system return function I is the sum of the stage return functions.

Let us now solve the problem by applying the dynamic programming technique. To do this, we define the *optimal return functions* $f_N[i]$ as

    $f_N[i] =$ maximum overall profit over N years,
        if we start with an i-year-old machine.

Using the principle of optimality, we have

$$f_N[i] = \max \begin{cases} R[i, K] + f_{N-1}\{G[i, K]\} \\ R[i, R] + f_{N-1}\{G[i, R]\} \end{cases}$$

$$= \max \begin{cases} p(i) - m(i) + f_{N-1}[i + 1] \\ p(0) - m(0) + s(i) - p + f_{N-1}[1] \end{cases} \tag{10.7}$$

$$f_1[i] = \max \begin{cases} p(i) - m(i) \\ p(0) - m(0) + s(i) - p \end{cases} \tag{10.8}$$

Now, Eqs. (10.7) and (10.8) can be solved recursively. We illustrate this with the help of a numerical example. (To simplify the calculations, we assume that the machine cannot be used for more than 10 years.) Suppose $p(i)$, $m(i)$, and $s(i) - p$ have the following values:

i	0	1	2	3	4	5	6	7	8	9	10
$p(i)$	60	55	50	45	40	37	35	33	30	20	10
$m(i)$	4	4	5	6	7	10	12	15	15	16	16
$s(i) - p$	0	−20	−30	−35	−50	−60	−80	−80	−90	−90	−100

Putting these values of $p(i)$, $m(i)$, and $s(i) - p$ in (10.7) and (10.8), we get the values of $f_n[i]$:

$i \to$	0	1	2	3	4	5	6	7	8	9	10
$f_1[i]_R^K$	56	51	45	39	33	27	23	18	15	4	
	56	36	26	21	6	−4	−24	−24	−34	−34	−44
$f_2[i]_R^K$	107	96	84	72	60	50	41	33	19	−40	
	107	87	77	72	57	47	37	27	17	17	7
$f_3[i]_R^K$	152	135	117	99	83	68	56	37	32	11	
	152	132	122	117	102	92	72	72	62	62	52
$f_4[i]_R^K$	191	173	162	141	125	99	95	80	77	56	
	191	171	161	156	141	131	111	111	101	101	91
$f_5[i]_R^K$	229	213	201	180	164	138	134	119	116	95	
	229	209	199	194	179	169	149	149	139	139	129
$f_6[i]_R^K$	269	252	239	218	202	176	172	157	154	133	
	269	249	239	234	219	209	189	189	179	179	169

These values may be used to calculate the *optimal replacement policy* up to 6 periods. The various 6-period optimal policies are as follows:

i	Optimal replacement policy
0	K(or R), K, K, R, K, K
1	K, K, R, K, K, K
2	K, R, K, R, K, K
3–10	R, K, K, R, K, K

BIBLIOGRAPHY

Angel, E., Dynamic Programming and Partial Differential Equations, Academic Press, New York, 1972.

Bellman, R., Dynamic Programming, Princeton University Press, Princeton, 1957.

Bellman, R. and Dreyfus, S. E., Applied Dynamic Programming, Princeton University Press, Princeton, 1962.

Jacobs, O. L. R., An Introduction to Dynamic Programming, Chapman and Hall, London, 1967.

Kaufmann, A. and Cruon, R., Dynamic Programming: Sequential scientific management, Academic Press, New York, 1967.

Nemhauser, G. L., Introduction to Dynamic Programming, Wiley, New York, 1966.

Roberts, S. M., Dynamic Programming in Chemical Engineering and Process Control, Academic Press, New York, 1964.

White, D. J., Dynamic Programming, Oliver and Boyd, Edinburgh, 1970.

PART IV STRUCTURAL APPLICATIONS

11

Optimal Design of Thin-Walled Columns under Axial Compressive Loading

11.1 INTRODUCTION

In aircraft structures and other sheet metal construction, stiffeners which are normally channel and Z-section are used. The behaviour of these sections as individual elements differs from that of the sheet-stiffener combination. Without precisely analyzing the sections as individual elements, we cannot understand the behaviour of the sheet-stiffener combination. The method proposed in this chapter aims at obtaining the minimum weight of the channel and Z-section columns under axial compressive loading by optimally proportioning the dimensions. To arrive at the optimal solution, we shall apply the *unconstrained optimization technique* using the interior penalty function approach (see Chapter 8).

11.2 FORMULATION

Let us assume that a column of given length is designed (as in Fig. 11.1) for a specified load;

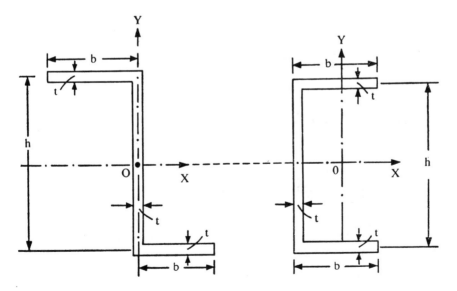

Fig. 11.1 *Z*- and Channel Section.

that the design variables for both the Z- and channel section are of flange width b, thickness t, and overall depth h; that the material used for the column is aluminium alloy with compressive yield stresses 25 ksi and 40 ksi, and the stress-strain relation is linear within the elastic limit. Further, let the behaviour constraints be imposed on different elastic buckling modes. Since the length of both the column and material is specified, optimizing the weight amounts to optimizing the cross-sectional area. The column shown in Fig. 11.1 has its ends simply-supported such that they are free to warp and rotate about the X- and Y-axis but cannot rotate about the Z-axis or deflect in the X- and Y-direction. Since the flange and web thicknesses are assumed to be the same for both sections, the design vector X contains three variables:

$$X^T = \{x_1, x_2, x_3\}, \tag{11.1}$$

where

$$x_1 = b, \qquad x_2 = t, \qquad x_3 = c = h/b. \tag{11.2}$$

11.3 OBJECTIVE FUNCTION

The objective function of our problem is to minimize the structural weight W which is defined by

$$W = \rho A_c L, \tag{11.3}$$

where ρ is the density of the material, A_c the area of cross-section of the column, and L the length of the column. For both the channel and Z-section, a fairly accurate expression for the area of cross-section is

$$A = 2bt + ht = bt(2 + h/b) = bt(2 + c). \tag{11.4}$$

Taking into consideration Eqs. (11.2), we may rewrite Eq. (11.4) as

$$W = x_1 x_2 (2 + x_3) \rho L. \tag{11.5}$$

Since ρ and L are constants, the objective function may be expressed as

$$f(X) = x_1 x_2 (2 + x_3). \tag{11.6}$$

11.4 BEHAVIOUR CONSTRAINTS

We now impose on the column behaviour five distinct restrictions which act as constraints in the choice of design variables. For any feasible design, the applied axial load P must not be greater than

    (i) the load $P_1(X)$ for which overall buckling of the column occurs;
    (ii) the load $P_2(X)$ for which local buckling of the flange takes place;
    (iii) the load $P_3(X)$ for which local buckling of the web is observed;
    (iv) the load $P_4(X)$ for which there is torsional buckling of the column;
    (v) the load $P_5(X)$ for which the yield stress is reached in the column.

A feasible design is obtained whenever

$$g_i(X) = P_i(X) - P \geqslant 0 \qquad (i = 1, 2, \ldots, 5). \tag{11.7}$$

The expressions for $P_i(X)$ can be obtained by the methods described in Chapters 2 and 3. Since the column is of uniform section, we can obtain a closed-form solution for $P_i(X)$.

Primary or Overall Buckling

We obtained primary or overall buckling (in Chapter 2) as

$$P_1 = \frac{\pi^2 E I_{\min}}{L^2}.$$ (11.8)

(i) For the Z-section, we get

$$I_{\min} = \tfrac{1}{2}[\tfrac{2}{3}b^3 t + \tfrac{1}{12}th^3 - \frac{t}{12}(h^6 + 36b^2h^4 + 64b^6 + 12bh^5 + 48b^4h^2 - 16b^3h^3)^{1/2} + \frac{bth^2}{2}].$$ (11.9)

Using Eqs. (11.2) and (11.8), we obtain

$$P_1(X) = \frac{\pi^2 E}{24L^2} x_1^3 x_2[(x_3^3 + 6x_3^2 + 8) - (x_3^6 + 12x_3^5 + 36x_3^4 - 16x_3^3 + 48x_3^2 + 64)^{1/2}].$$ (11.10)

(ii) For the channel section, we have

$$P_1(X) = \frac{\pi^2 E}{3L^2} x_1^3 x_2 (\frac{1 + 2x_3}{2 + x_3}).$$ (11.11)

Local Buckling (of flanges)

In evaluating the local buckling stress of a flange, we assume that the flange is simply-supported at the edge where it joins the web and is free at the other edge. The corresponding buckling stress formula is

$$\sigma_{cr} = \frac{k\pi^2 E}{12(1 - v^2)}(\frac{t}{b})^2,$$ (11.12)

where $k = 0.43$ (see Gerard and Becker[1]). For both the channel and Z-section, the local buckling stresses of the flange are the same. Using Eqs. (11.2), we get the buckling load

$$P_2(X) = \sigma_{cr} A_c = \frac{(0.0358)\pi^2 E}{(1 - v^2)} \frac{x_2^3}{x_1} (2 + x_3).$$ (11.13)

Local Buckling (of webs)

We shall treat the web as a plate which is simply-supported on all sides. If, however, the edge that connects the web and the flange in the manner of an elastic support is considered, a more accurate estimate of the buckling load is possible. Here, the coefficient k in the buckling formula is equal to 4 (see Gerard and Becker[1]). Using this, we get the critical stress

$$\sigma_{cr} = \frac{4\pi^2 E}{12(1 - v^2)}(\frac{t}{h})^2$$ (11.14)

and the buckling load

$$P_3(X) = \frac{\pi^2 E}{3(1 - v^2)} \frac{x_2^3}{x_1} \frac{(2 + x_3)}{x_3^2}.$$ (11.15)

Torsional Buckling

Z-section   This section being doubly symmetric, $x_0 = 0$, $y_0 = 0$ (see Section 2.6). Consequently, the flexure and torsional modes of the column are uncoupled. The possible buckling stresses are

$$\sigma_{cr\,1} = \frac{\pi^2 EI_{min}}{L^2}, \tag{11.16}$$

$$\sigma_{cr\,2} = \frac{\pi^2 EI_{max}}{L^2}, \tag{11.17}$$

$$\sigma_{cr\,3} = \frac{E\Gamma\pi^2/L^2 + GJ}{I_p}. \tag{11.18}$$

Equation (11.16) is already contained in Eq. (11.8). Hence, we consider (11.18) for torsional buckling of the column, leading to the buckling load

$$P_4(X) = \frac{\pi^2 E}{L^2} \frac{x_1^3 x_2 x_3^2 (1 + 2x_3)}{8 + 6x_3^2 + x_3^3} + \frac{4Gx_2^3 (2 + x_3)^2}{x_1(8 + 6x_3^2 + x_3^3)}. \tag{11.19}$$

(The calculation of Γ and J is similar to that given in Section 2.5.)

Channel section   This is a singly symmetric section, the X-axis being the axis of symmetry. Hence, $y_0 = 0$. In this case, there is coupling between one of the flexure modes and the torsional mode. Since the combined equation has two roots, we take the lower value as the critical stress and get the buckling load

$$\begin{aligned}
P_4(X) = {} & \frac{\pi^2 E x_1^3 x_2 x_3^2}{24L^2} \Big(\frac{x_3^5 + 14x_3^4 + 60x_3^3 + 104x_3^2 + 232x_3 + 336}{x_3^4 + 8x_3^3 + 12x_3^2 + 8x_3 + 4}\Big) \\
& + \frac{2Gx_2^3 (2 + x_3)^3}{x_1(x_3^4 + 8x_3^3 + 12x_3^2 + 8x_3 + 4)} \\
& - [\{\frac{\pi^2 E x_1^3 x_2 x_3^2}{24L^2}\Big(\frac{288 + 128x_3 - 56x_3^2 - 60x_3^3 - 14x_3^4 - x_3^5}{x_3^4 + 8x_3^3 + 12x_3^2 + 8x_3 + 4}\Big) \\
& + \frac{2\Gamma x_2^3 (2 + x_3)^3}{x_1(x_3^4 + 8x_3^3 + 12x_3^2 + 8x_3 + 4)}\}^2 + \frac{(1.33)\pi^4 E^2 x_1^6 x_3^4 x_2^2 (3 + x_3)^2}{L^4 (x_3^4 + 8x_3^3 + 12x_3^2 + 8x_3 + 4)}]^{1/2},
\end{aligned} \tag{11.20}$$

where

$$\Gamma = \frac{x_1^5 x_2 x_3^2}{12} \frac{2x_3^2 + 15x_3 + 26}{(2 + x_3)^2}. \tag{11.21}$$

Material Yield Stress

In all the constraints so far considered, no limit has been placed on the critical stress. If the stresses exceed the elastic range, Eqs. (11.8)–(11.21) are not applicable. However, if we limit the stresses in the column by including a yield constraint, the stresses in the buckling mode would lie within the elastic range. This constraint imposes a lower bound on the cross-sectional area, giving the buckling load

$$P_5(X) = \sigma_y x_1 x_2 (2 + x_3), \tag{11.22}$$

where σ_y is the yield stress in compression.

11.5 SIDE CONSTRAINTS

The design variables must be positive in order to be realizable. In structural design, the design variables are normally bounded and depend on their availability and the handling restrictions. In our problem, the variables should be greater than zero since some or all of the behaviour constraints would be violated if even one of the variables is zero. The non-negativity restrictions on the design variables are

$$x_i \geqslant 0 \quad (i = 1, 2, 3). \tag{11.23}$$

The formulation presented in Sections 11.2–11.4 may be stated in mathematical programming language as follows. For the Z- and channel section, find X^* to

$$\text{minimize } f(X) = x_1 x_2 (2 + x_3) \tag{11.24}$$

subject to

$$
\begin{aligned}
g_1(X) &= P_1(X) - P \geqslant 0, \\
g_2(X) &= P_2(X) - P \geqslant 0, \\
g_3(X) &= P_3(X) - P \geqslant 0, \\
g_4(X) &= P_4(X) - P \geqslant 0, \\
g_5(X) &= P_5(X) - P \geqslant 0, \\
x_i &\geqslant 0 \quad (i = 1, 2, 3).
\end{aligned}
\tag{11.25}
$$

The expressions for $P_1(X), P_2(X), \ldots, P_5(X)$ are different for the channel and Z-section.

11.6 METHOD OF SOLUTION

We use the sequential unconstrained minimization technique (SUMT) (see Fiacco and McCormick[2,3,4]) to reach the optimal solution. In so doing, a new function, called the P-function, is formed by augmenting the objective function $f(X)$ with the constraints; this is done by using a *penalty factor r* such that

$$P_f(X, r) = f(X) + r \sum_{i=1}^{m} \frac{1}{g_i(X)}. \tag{11.26}$$

The variable metric method of Fletcher and Powell[5] is employed for minimization. This method requires only the knowledge of first partial derivatives of the functions. For the two columns considered, the P-function is

$$P_f(X, r) = x_1 x_2 (2 + x_3) + r \left(\frac{1}{x_1} + \frac{1}{x_2} + \frac{1}{x_3} + \frac{1}{g_1(X)} + \frac{1}{g_2(X)} + \frac{1}{g_3(X)} + \frac{1}{g_4(X)} + \frac{1}{g_5(X)} \right). \tag{11.27}$$

In minimizing a function by using the Fletcher-Powell method, the following steps are involved.

Step 1   Let X^0 be the starting point and $[H^0]$ an arbitrary positive definite matrix of order $n \times n$, where n is the number of design variables. Usually, $[H^0]$ is an $n \times n$ identity matrix.

Step 2   Find the gradient at X^0, i.e., $\nabla f(X^0)$. For the i-th iteration X^i, the gradient is $\nabla f(X^i)$ and the positive definite matrix is $[H^i]$.

Step 3   Let

$$S^i = -[H^i]\nabla f(X^i),\tag{11.28}$$

$$\mu_i = \lambda^i S^i,\tag{11.29}$$

$$X^{i+1} = X^i + \mu_i,\tag{11.30}$$

where λ^i is a scalar greater than 0, chosen to minimize $f(X)$ along S^i, the starting point being X^i.

Step 4   Evaluate

$$Y^i = \nabla f(X^{i+1}) - \nabla f(X^i).\tag{11.31}$$

Step 5   The new Hessian matrix is given by

$$[H^{i+1}] = [H^i] + \frac{(\mu^i)(\mu^i)^T}{(\mu^i)^T Y^i} - \frac{[H^i](Y^i)(Y^i)^T[H^i]^T}{(Y^i)^T[H^i](Y^i)}.\tag{11.32}$$

Step 6   Return to step 3.

Step 7   Stop the algorithm when the convergence criteria are satisfied.

The convergence criteria applied in our study are: (i) there is no change in the P-function value for two consecutive iterations when the step length in a direction of movement is zero; (ii) the gradient vector corresponding to the minimum is checked for a very low value of $0(10^{-3})$; (iii) about 15 P-function minimizations are carried out when the final value of the *penalty factor r* tends to zero.

The flowchart used for developing the program on IBM 7044 is depicted in Fig. 11.2. The results for both the Z- and channel section for the following set of data are presented in Tables 11.1–11.4:

$$P = 1000 \text{ lb}, \qquad L = 25 \text{ in}, \qquad \sigma_y = 25,000 \text{ psi}, \qquad E = 10^7 \text{ psi}, \qquad \nu = 0.3.$$

In order to determine the nature of the optimum point, that is, whether it is global or local, the sequential minimization algorithm of the P-function is carried out from different initial values.

11.7   RESULTS AND DISCUSSION

Tables 11.1–11.3 give the results for the Z-section with different starting points. It is evident that, in all the three cases, the final point remains the same up to four decimal accuracy which indicates that the global optimum is attained. Besides, the primary and local buckling modes of the flange are the critical modes at the optimum point. Another interesting result is that the optimum h/b ratio for the Z-section is the same for all designs, regardless of the design considerations.

Table 11.4 gives the values obtained for the channel section. Here, primary and torsional buckling modes are active at the optimal point; the h/b ratio remains constant but is different from that of the Z-section.

This study indicates that, for the optimum design, all the failure modes do not always occur simultaneously.

Main Program

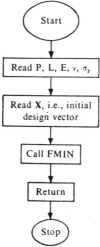

Subroutine FMIN

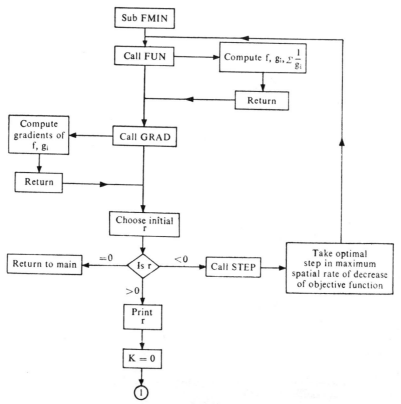

Fig. 11.2 Flowchart for Optimization under Compressive Loading (cont.).

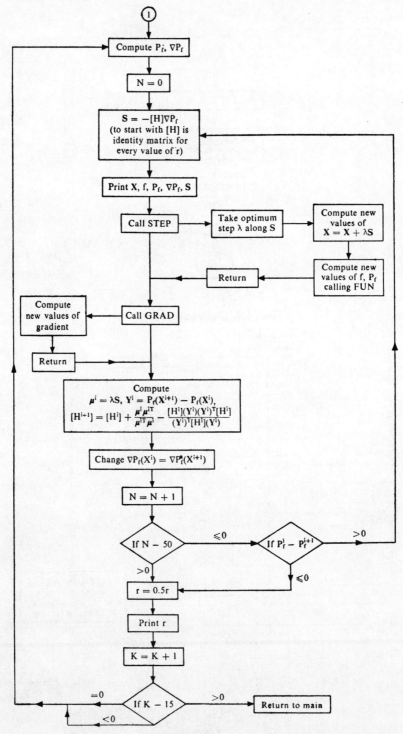

Fig. 11.2   Flowchart for Optimization under Compressive Loading (cont.).

Subroutine STEP

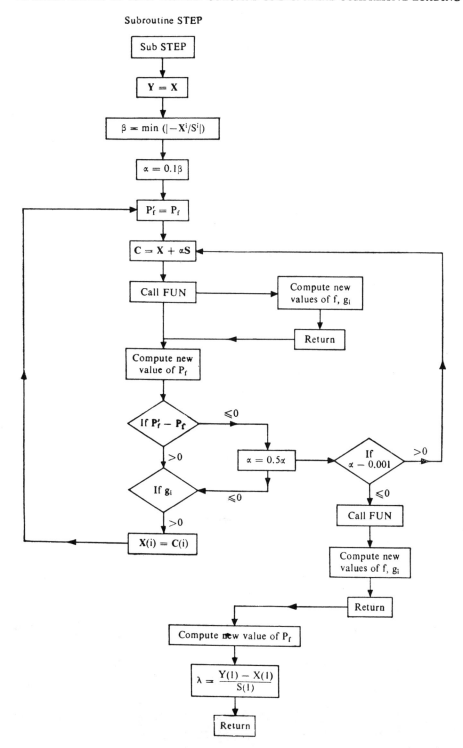

Fig. 11.2 Flowchart for Optimization under Compressive Loading.

Table 11.1   Optimal Values of x_1, x_2, and x_3 (for Different Values of r) for Columns with Z-Section

Given

$$P = 1000 \text{ lb}, \quad L = 25 \text{ in}, \quad \sigma_y = 25 \text{ ksi}, \quad E = 10^7 \text{ psi}, \quad \nu = 0.3.$$

Assume

$$x_1 = 1.5, \quad x_2 = 0.05, \quad x_3 = 1.5, \quad r_1 = 0.0130057.$$

Initial values

$$f(X) = 0.2625, \quad g_1(X) = 4.336 \times 10^3, \quad g_2(X) = 1.335 \times 10^2,$$
$$g_3(X) = 3.686 \times 10^3, \quad g_4(X) = 9.272 \times 10^3, \quad g_5(X) = 5.562 \times 10^3.$$

r	x_1	x_2	x_3	$f(X)$	$P_f(X)$	$g_1(X)$	$g_2(X)$	$g_3(X)$	$g_4(X)$	$g_5(X)$
1.3×10^{-2}	0.7553	0.08684	1.292	0.21597	0.35469	14.103	10,093	60,795	7918	4399
6.5×10^{-3}	0.8653	0.05791	1.286	0.16468	0.29008	11.215	1866	15,115	2605	3117
3.25×10^{-3}	0.9502	0.04230	1.322	0.13353	0.21674	8.575	28.79	4475	1406	2338
1.625×10^{-3}	0.9502	0.04230	1.322	0.13353	0.16822	8.575	28.79	4475	1406	2338
8.13×10^{-4}	0.9502	0.04230	1.322	0.13353	0.15088	8.575	28.79	4475	1406	2338
4.06×10^{-4}	0.9502	0.04230	1.322	0.13353	0.14221	8.575	28.79	4475	1406	2338
2.03×10^{-4}	0.9502	0.04230	1.322	0.13353	0.13787	8.575	28.79	4475	1406	2338
1.02×10^{-4}	0.9083	0.04029	1.580	0.13101	0.13383	1.397	2.003	2733	1366	2275
5.1×10^{-5}	0.9083	0.04029	1.580	0.13101	0.13209	1.397	2.0025	2733	1366	2275
2.5×10^{-5}	0.9083	0.04029	1.580	0.13101	0.13155	1.397	2.0025	2733	1366	2275
1.3×10^{-5}	0.9083	0.04029	1.580	0.13101	0.13128	1.397	2.0025	2733	1366	2275
6.0×10^{-6}	0.9069	0.04021	1.589	0.13091	0.13111	0.352	0.4902	2683	1363	2273
3.0×10^{-6}	0.9069	0.04021	1.589	0.13091	0.13098	0.3517	0.4902	2683	1363	2273
2.0×10^{-6}	0.9069	0.04021	1.589	0.13091	0.13093	0.3517	0.4902	2683	1363	2273
1.0×10^{-6}	0.9068	0.04021	1.590	0.13089	0.13092	0.1251	0.1673	2681	1362	2272

Table 11.2 Optimal Values of x_1, x_2, and x_3 (for Different Values of r) for Columns with Z-Section

Given

$$P = 1000 \text{ lb}, \quad L = 25 \text{ in}, \quad \sigma_y = 25 \text{ ksi}, \quad E = 10^7 \text{ psi}, \quad \nu = 0.3.$$

Assume

$$x_1 = 1.75, \quad x_2 = 0.07, \quad x_3 = 1.9, \quad r_1 = 0.03343821.$$

Initial values

$$f(X) = 0.47775, \quad g_1(X) = 13{,}428, \quad g_2(X) = 1970,$$
$$g_3(X) = 6655, \quad g_4(X) = 28{,}366, \quad g_5(X) = 10{,}944.$$

r	x_1	x_2	x_3	$f(X)$	$P_f(X)$	$g_1(X)$	$g_2(X)$	$g_3(X)$	$g_4(X)$	$g_5(X)$
3.344×10^{-2}	0.6313	0.1490	1.2975	0.3102	0.6150	19.933	66,181	3,70,186	43,994	6753
1.672×10^{-2}	0.7195	0.1004	1.2943	0.2380	0.4418	15.397	17,031	99,125	12,386	4952
8.359×10^{-3}	0.8235	0.0671	1.2883	0.1818	0.3236	12.166	3693	25,301	3788	3544
4.179×10^{-3}	0.8235	0.0671	1.2883	0.1818	0.2461	12.166	3693	25,301	3788	3544
2.090×10^{-3}	0.8235	0.0671	1.2883	0.1818	0.2139	12.166	3693	25,301	3788	3544
1.045×10^{-3}	0.8235	0.0671	1.2883	0.1818	0.1978	12.166	3693	25,301	3788	3544
5.22×10^{-4}	0.9139	0.04056	1.5396	0.1312	0.1453	3.206	4.831	2943	1371	2280
2.61×10^{-4}	0.9139	0.04056	1.5396	0.1312	0.1352	3.206	4.831	2943	1371	2280
1.31×10^{-4}	0.9139	0.04056	1.5396	0.1312	0.1332	3.206	4.831	2943	1371	2280
6.5×10^{-5}	0.9139	0.04056	1.5396	0.1312	0.1322	3.206	4.831	2943	1371	2280
3.3×10^{-5}	0.9139	0.04056	1.5396	0.1312	0.1317	3.206	4.831	2943	1371	2280
1.6×10^{-5}	0.9139	0.04056	1.5396	0.1312	0.1314	3.206	4.831	2943	1371	2280
8×10^{-6}	0.9069	0.04021	1.5894	0.13092	0.13098	0.3794	0.5604	2684	1364	2273

Table 11.3 Optimal Values of x_1, x_2, and x_3 (for Different Values of r) for Columns with Z-Section

Given

$$P = 1000 \text{ lb}, \quad L = 25 \text{ in}, \quad \sigma_y = 25 \text{ ksi}, \quad E = 10^7 \text{ psi}, \quad \nu = 0.3.$$

Assume

$$x_1 = 1.3, \quad x_2 = 0.048, \quad x_3 = 1.6, \quad r_1 = 0.01073138.$$

Initial values

$$f(X) = 0.22464, \quad g_1(X) = 2537, \quad g_2(X) = 190.22,$$
$$g_3(X) = 3325, \quad g_4(X) = 6139, \quad g_5(X) = 4616.$$

r	x_1	x_2	x_3	$f(X)$	$P_f(X)$	$g_1(X)$	$g_2(X)$	$g_3(X)$	$g_4(X)$	$g_5(X)$
1.073×10^{-2}	0.78415	0.07767	1.2906	0.20042	0.36140	13.216	6642	41,681	5693	4010
5.366×10^{-3}	0.78028	0.07764	1.2936	0.19953	0.29103	0.4725	6678	41,686	5685	3988
2.683×10^{-3}	0.94283	0.04819	1.3620	0.13280	0.20215	7.6793	18.960	4109	1395	2320
1.341×10^{-3}	0.94283	0.04819	1.3620	0.13280	0.16262	7.6793	18.960	4109	1395	2320
6.707×10^{-4}	0.94283	0.04819	1.3620	0.13280	0.14771	7.6793	18.960	4109	1395	2320
3.354×10^{-4}	0.94283	0.04819	1.3620	0.13280	0.14025	7.6793	18.960	4109	1395	2320
1.677×10^{-4}	0.94283	0.04819	1.3620	0.13280	0.13652	7.6793	18.960	4109	1395	2320
8.384×10^{-5}	0.90802	0.04028	1.5816	0.13099	0.13333	1.2704	1.8171	2725	1365	2275
4.192×10^{-5}	0.90802	0.04028	1.5816	0.13099	0.13193	1.2704	1.8171	2725	1365	2275
2.096×10^{-5}	0.90802	0.04028	1.5816	0.13099	0.13146	1.2704	1.8171	2725	1365	2275
1.048×10^{-5}	0.90802	0.04028	1.5816	0.13099	0.13123	1.2704	1.8171	2725	1365	2275
5.240×10^{-6}	0.90686	0.04021	1.5896	0.13091	0.13107	0.31407	0.4463	2682	1363	2272
2.620×10^{-6}	0.90686	0.04021	1.5896	0.13091	0.13096	0.31407	0.4463	2682	1363	2272
1.310×10^{-6}	0.90686	0.04021	1.5896	0.13091	0.13094	0.31407	0.4463	2682	1363	2272
6.550×10^{-7}	0.90683	0.04021	1.5897	0.13089	0.13092	0.11100	0.14671	2681	1362.9	2272

Table 11.4 Optimal Values of x_1, x_2, and x_3 (for Different Values of r) for Columns with Channel Section

Given

$P = 1000$ lb, $L = 25$ in, $\sigma_y = 25$ ksi, $E = 10^7$ psi, $\nu = 0.3$.

Assume

$x_1 = 2.0$, $x_2 = 0.1$, $x_3 = 1.0$, $r_1 = 0.060015$.

Initial values

$f(X) = 0.6000$, $g_1(X) = 41{,}111$, $g_2(X) = 4830$,

$g_3(X) = 53{,}229$, $g_4(X) = 8695$, $g_5(X) = 14{,}000$.

r	x_1	x_2	x_3	$f(X)$	$P_f(X)$	$g_1(X)$	$g_2(X)$	$g_3(X)$	$g_4(X)$	$g_5(X)$
6.002×10^{-2}	0.4493	0.2381	0.7426	0.28342	0.76305	30.148	3,19,295	54,01,915	49.881	6336
3.001×10^{-2}	0.4998	0.1700	0.7675	0.23519	0.51241	23.682	1,04,777	16,69,514	91.087	4880
1.500×10^{-2}	0.5655	0.1166	0.7719	0.18275	0.35858	18.636	29,180	4,70,143	42.967	3569
7.502×10^{-3}	0.5655	0.1166	0.7719	0.18275	0.26899	18.636	29,180	4,70,143	42.967	3569
3.751×10^{-3}	0.6687	0.0659	0.9638	0.13062	0.19753	24.56	3932	48,394	10.837	2265
1.875×10^{-3}	0.6687	0.0659	0.9638	0.13062	0.15219	24.56	3932	48,394	10.837	2265
9.377×10^{-4}	0.6687	0.0659	0.9638	0.13062	0.14140	24.56	3932	48,394	10.837	2265
4.689×10^{-4}	0.6687	0.0659	0.9638	0.13062	0.13601	24.56	3932	48,394	10.837	2265
2.344×10^{-4}	0.6687	0.0659	0.9638	0.13062	0.13371	24.56	3932	48,394	10.837	2265
1.172×10^{-4}	0.6687	0.0659	0.9638	0.13062	0.13196	24.56	3932	48,394	10.837	2265
5.861×10^{-5}	0.6995	0.04784	1.5782	0.11975	0.12118	1.352	1177	7129	1.1414	1994
2.930×10^{-5}	0.6995	0.04784	1.5782	0.11975	0.12043	1.352	1177	7129	1.1414	1994
1.465×10^{-5}	0.6995	0.04784	1.5782	0.11975	0.12009	1.352	1177	7129	1.1414	1994
7.326×10^{-6}	0.7014	0.04681	1.6450	0.11967	0.11987	0.53535	1072	6121	0.3948	1992
3.663×10^{-6}	0.7014	0.04681	1.6450	0.11967	0.11975	0.53535	1072	6121	0.3948	1992
1.832×10^{-6}	0.7014	0.04681	1.6450	0.11967	0.11971	0.53535	1072	6121	0.3948	1992
9.158×10^{-7}	0.7013	0.04680	1.6452	0.11965	0.11966	0.18214	1071	6117	0.1307	1991

REFERENCES

1 Gerard, G. and Becker, H., Handbook of Structural Stability, Part I—Buckling of flat plates, NACA, TN 3781, 1957.

2 Fiacco, A. V. and McCormick, G. P., The sequential unconstrained minimization technique for nonlinear programming: a primal dual method, *Management Sc.*, **10**, 2, 360–66, 1964.

3 Fiacco, A. V. and McCormick, G. P., Computational algorithm for the sequential unconstrained minimization technique for nonlinear programming, *Management Sc.*, **10**, 4, 601–17, 1964.

4 Fiacco, A. V. and McCormick, G. P., Nonlinear Programming: Sequential unconstrained minimization technique, Wiley, New York, 1968.

5 Fletcher, R. and Powell, M. J. D., A rapidly convergent descent method for minimization, *Computer J.*, **6**, 2, 163–68, 1963.

12

Minimum Weight Design of Multicell Box Beams in Bending

12.1 INTRODUCTION

Aircraft wings must have the required strength and stiffness in addition to the minimum weight. One of the simplest designs for these wings is the multiweb box beam which is made up of a thick skin and a number of relatively thin webs suitably located in the spanwise direction. The aerodynamic characteristics determine the external shape and other dimensions of the wing. However, the structural designer has considerable freedom in the choice of a structural configuration.

For a multiweb box beam subjected to a pure bending load, the primary purpose of the webs is to displace the effective area as far as possible from the neutral axis and divide the compression cover into a series of panels such that the buckling stress for each panel is quite high. This problem has been investigated by several workers (see Gerard and Becker[1], Rosen[2], Needham[3], Gerard and Becker[4], Krishnan and Ramachandra[5]) and, in our study, it is formulated as an NLP problem and solved by using the sequential unconstrained minimization technique (SUMT).

12.2 FORMULATION

Figure 12.1a shows the normal cross-section of a multiweb box beam. For our analysis, it is idealized as a rectangular box beam (see Fig. 12.1b). Since the wing is designed to carry only the bending load, the webs are idealized as an *I*-section (see Fig.12.2). The *equivalent height h* of the idealized section is determined by keeping the overall second moment of area and the chord *c* of the idealized wing the same as those of the actual wing.

We shall consider two failure modes, namely, flexure failure and local crippling of the web flange and skin. It has been shown experimentally by Needham[6] and Pride and Anderson[7] that failure modes such as web buckling and failure of the webs normal to the plane of the skin are not generally encountered in a well-proportioned box beam.

We shall assume in our analysis that
    (i) the stresses are within the elastic limit;
    (ii) the effective bending elements of the web are equal angles with the leg width *a*;
    (iii) the effective bending elements involve twice the effective width of the skin;
    (iv) all vertical webs are of equal thickness.

Thus, the optimization problem has three design variables, namely, skin thickness t_s, web thickness t_w, and number of webs n_w. Hence, the design vector X is given by

$$X^T = \{t_s, t_w, n_w\} \tag{12.1}$$
$$= \{x_1, x_2, x_3\}. \tag{12.2}$$

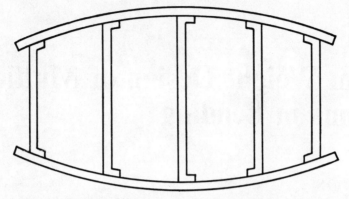

(a) Cross-section of multiweb box beam

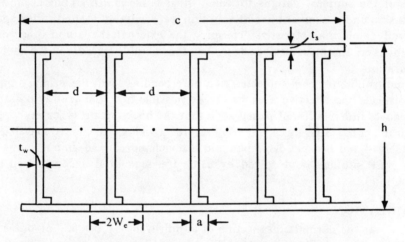

(b) Idealized cross-section of multiweb box beam

Fig. 12.1   Actual and Idealized Cross-Sections of Multiweb Box Beam.

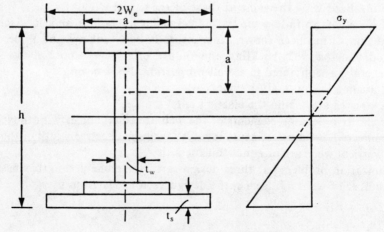

Fig. 12.2   Idealization of Web.

12.3 OBJECTIVE FUNCTION

The objective function $f(X)$ is described by

$$f(X) = \rho A L. \tag{12.3}$$

Since ρ and L are constants, minimizing the weight of the structure amounts to minimizing the area of the cross-section which is given by

$$A = 2ct_s + t_w n_w (h - 2t_s - 2t_w + 2a). \tag{12.4}$$

In view of Eq. (12.4), the objective function becomes

$$f(X) = 2cx_1 + x_2 x_3 (h - 2x_1 - 2x_2 + 2a). \tag{12.5}$$

12.4 BEHAVIOUR CONSTRAINTS

When both the failure modes occur simultaneously, the design is optimum. Thus,

$$\begin{aligned} g_1(X) &= \sigma_b - \alpha\sigma_y \geqslant 0, \\ g_2(X) &= \sigma_{cc} - \alpha\sigma_y \geqslant 0, \end{aligned} \tag{12.6}$$

where σ_b is the bending stress, σ_{cc} the crippling stress, and σ_y the yield stress. From Eqs. (12.6), we derive

$$\begin{aligned} \sigma_b - \alpha\sigma_y &\geqslant 0, \\ \frac{Mh}{2I} - \alpha\sigma_y &\geqslant 0. \end{aligned} \tag{12.7}$$

The effective width W_e is given by

$$\begin{aligned} W_e &= (0.85)t_s\sqrt{\eta\sqrt{E/\sigma_{cc}}}, \\ W_e &= Dt_s, \end{aligned} \tag{12.8}$$

where

$$D = (0.85)\sqrt{\eta\sqrt{E/\sigma_{cc}}}.$$

The plasticity reduction factor η as given by Stowell[8] is

$$\eta = (\frac{E_s}{E})[\tfrac{1}{2} + \tfrac{1}{2}(\tfrac{1}{4} + \tfrac{3}{4}\frac{E_t}{E_s})^{1/2}], \tag{12.9}$$

where E_s and E_t are the secant and the tangent modulus, respectively. For a well-proportioned box structure made of aluminium alloy, $E_s/E = 0.7$ and $E_t/E_s = 0.2$. Using W_e, we can write the constraint function $g_1(X)$ as

$$\begin{aligned} g_1(X) &= \frac{D}{3}x_1^4 x_3 + Dx_1^2 x_3(h - x_1)^2 + \frac{ax_3}{12}(h - 2x_1)^3 - \frac{x_3 x_2}{12}(h - 2x_1 - 2a)^3 \\ &\quad - \frac{x_3(a - x_2)}{12}(h - 2x_1 - 2x_3)^3 - \frac{Mh}{2\alpha\sigma_y} \geqslant 0. \end{aligned} \tag{12.10}$$

The crippling stress σ_{cc} for the compression element of the web flange is obtained from the empirical equation (due to Needham[6])

$$\sigma_{cc} = C_e\sqrt{E\sigma_y}(\frac{t_w}{b})^{0.75}, \tag{12.11}$$

where C_e is the constant, depending on the support condition, and

$$b' = \tfrac{1}{2}(2a - t_w - t_s).$$

Substituting for σ_{cc} from Eqs. (12.6) in Eq. (12.11), we have

$$t_w = K(2a - t_w - t_s), \tag{12.12a}$$
$$x_2 = K(2a - x_2 - x_1), \tag{12.12b}$$

where

$$K = \tfrac{1}{2}(\frac{4C_e}{\alpha}E^2\sigma_y^2)^{-1/3}.$$

Equation (12.12b) may be rewritten as

$$g_2(X) = x_2 - R(2a - x_1) \geqslant 0, \tag{12.13}$$

where

$$R = \frac{K}{1 + K}.$$

12.5 SIDE CONSTRAINTS

In accordance with Section 11.5, the side constraints are here defined by

$$g_3(X) = x_1 - x_2 \geqslant 0, \tag{12.14}$$
$$g_4(X) = 50 - a/x_2 \geqslant 0, \tag{12.15}$$
$$g_5(X) = a/x_2 - 9 \geqslant 0, \tag{12.16}$$
$$g_6(X) = x_1 \geqslant 0, \tag{12.17}$$
$$g_7(X) = x_3 \geqslant 0. \tag{12.18}$$

These constraints are introduced in conformity with experiments conducted by Needham[3].

12.6 METHOD OF SOLUTION

To obtain the optimal solution, we use the SUMT by applying the penalty function approach. The penalty function $P_f(X)$ is given by

$$P_f(X) = 2cx_1 + x_2x_3(h - 2x_1 - 2x_2 + 2a)$$
$$+ r(\frac{1}{g_1(X)} + \frac{1}{g_2(X)} + \frac{1}{g_3(X)} + \frac{1}{g_4(X)} + \frac{1}{g_5(X)} + \frac{1}{g_6(X)} + \frac{1}{g_7(X)}). \tag{12.19}$$

12.7 RESULTS AND DISCUSSION

Using the data

$$M = 25,000 \text{ kg}, \quad a = 2 \text{ cm}, \quad c = 200 \text{ cm}, \quad h = 20 \text{ cm}$$

and the Fletcher-Powell variable metric method (discussed in Section 11.6) for minimization, we obtain the optimum design variables

$$x_1 = 0.2832, \quad x_2 = 0.1757, \quad x_3 = 18.878.$$

The cross-sectional area corresponding to these variables is equal to 189.8 cm².

Table 12.1 shows the results obtained from different starting design vectors. It should be noted that since there is a large number of minima, the minimum of the minima is picked up as the optimum value.

Table 12.1 Optimal Values of x_1, x_2, and x_3 for Different Initial Points

Given

$$M = 25{,}000 \text{ kg}, \qquad a = 2 \text{ cm}, \qquad c = 200 \text{ cm}, \qquad h = 20 \text{ cm}.$$

	Initial values						Final values					
x_1	x_2	x_3	$f(X)$	$P_f(X)$	r		x_1	x_2	x_3	$f(X)$	$P_f(X)$	r
0.45	0.19	18	257.7	268.7	0.2119		0.283	0.1757	17.87	189.8	189.8	10^{-38}
0.45	0.19	18	257.7	545×10^4	100		0.368	0.171	12.79	197.6	197.9	0.1×10^{-5}
0.55	0.199	10	264.7	0.6×10^3	10		0.406	0.170	10.91	205	205	0.1×10^{-5}
0.80	0.19	30	445	4749	1		0.340	0.173	14.40	193.6	193.6	0.1×10^{-5}
0.55	0.199	10	264.7	298	1		0.340	0.173	14.47	193	193.6	0.1×10^{-6}

REFERENCES

1 Gerard, G. and Becker, H., Handbook of Structural Stability, Part IV—Failure of plates and composite elements, NACA, TN 3784, 1957.

2 Rosen, W. B., Analysis of Ultimate Strength and Optimum Properties of Multiweb Wing Structures, NACA, TN 3633, 1956.

3 Needham, R. A., Ultimate strength of multiweb box beam under pure bending, *J. Aero. Sc.*, **22**, 10, 781–86, 1955.

4 Gerard, G. and Becker, H., Handbook of Structural Stability, Part VII—Strength of thin wing construction, NASA, TND 162, 1959.

5 Krishnan, S. and Ramachandra, H. N., Minimum weight design of multiweb box beam, *J. Aero. Soc. Ind.*, **21**, 3, 265–74, 1969.

6 Needham, R. A., The ultimate strength of aluminium alloy formed structural shapes in compression, *J. Aero. Sc.*, **21**, 4, 217–29, 1954.

7 Pride, R. A. and Anderson, M., Experimental Investigation of the Pure Bending Strength of 75S–T6 Aluminium Alloy Multiweb Beams with Formed Channel Webs, NACA, TN 3082, 1954.

8 Stowell, E. Z., A Unified Theory of the Plastic Buckling of Columns and Plates, NACA, TN 1556, 1948.

13

Minimum Weight Design of Multicell Wings for Strength and Frequency Requirements

13.1 INTRODUCTION

In Chapters 11 and 12, the minimum weight design of structures subjected to static constraints has been considered. In this chapter, a problem that involves constraints on natural frequency is analyzed. By properly choosing the wing dimensions of an aircraft, the first few frequencies can be adjusted to increase the flutter speed for a minimum wing weight. Our study is aimed at finding out the optimum number of cells for such a wing, with constraints on its strength and natural frequencies. To determine the minimum structural weight satisfying a specific requirement, Rubin[1] has employed gradient equations which express the rate of change of frequency with respect to the design variables. Employing the matrix method together with the feasible direction method, Fox and Kapoor[2] arrived at the optimum design of planar truss frames with constraints on their dynamic response and frequency characteristics.

In this study, the constrained problem is solved by means of the transformation method, developed by Box[3], and the penalty function approach, evolved by Fiacco and McCormick[4, 5]. These unconstrained methods are better suited than other sequential unconstrained minimization techniques (SUMT) (see Zoutendijk[6]) for purposes of computation.

13.2 FORMULATION

For a rectangular wing of uniform skin thickness and $N - 1$ spars, the mass M may be written as

$$M = \rho LC \left(k_0 t_s + \sum_{i=1}^{N-1} k_i t_i \right), \tag{13.1}$$

where

$N =$ number of cells,

$L =$ semispan of the wing,

$C =$ chord of the airfoil,

$t_s =$ skin thickness,

$t_i =$ thickness of the i-th spar,

$k_0 =$ perimeter of the airfoil for unit chord,

$k_i =$ height of the i-th spar as percentage chord,

$\rho =$ mass density (assuming wing and spars are of the same material).

The optimization problem is written as:

$$\text{Minimize } M(L, C, t_s, t_i) \tag{13.2}$$

subject to

$$\sigma_s \leqslant \sigma_y,$$

$$\omega_l \leqslant \omega \leqslant \omega_u,$$

$$L_l \leqslant L \leqslant L_u,$$

$$C_l \leqslant C \leqslant C_u, \tag{13.3}$$

$$t_{sl} \leqslant t_s \leqslant t_{su},$$

$$t_{il} \leqslant t_i \leqslant t_{iu} \qquad (i = 1, 2, \ldots, N-1),$$

where

$\omega =$ first natural frequency of the structure,

$\sigma_s =$ maximum stress imposed on the structure,

$l, u =$ lower and upper bounds, respectively, on the variables,

last four inequalities are side constraints.

By employing the *transformations* used by Box[3], we get

$$L = L_l + (L_u - L_l) \sin^2 d_1,$$

$$C = C_l + (C_u - C_l) \sin^2 d_2,$$

$$t_s = t_{sl} + (t_{su} - t_{sl}) \sin^2 d_3, \tag{13.4}$$

$$t_i = t_{il} + (t_{iu} - t_{il}) \sin^2 d_{3+i} \qquad (i = 1, 2, \ldots, N-1).$$

When the side constraints are removed, the problem is restated as:

$$\text{Minimize } M(X). \tag{13.5}$$

13.3 BEHAVIOUR CONSTRAINTS

The behaviour constraints are specified on the maximum stress and the first natural frequency of the structure as

$$g_1(X) = \sigma_s - \sigma_y \leqslant 0 \equiv \sigma_s - 30,000 \leqslant 0,$$

$$g_2(X) = \omega_l - \omega \leqslant 0 \equiv \omega - 35 \leqslant 0, \tag{13.6}$$

$$g_3(X) = \omega - \omega_u \leqslant 0 \equiv 28 - \omega \leqslant 0,$$

where X is the vector of the transformed design variables x_i $(i = 1, 2, \ldots, N+2)$.

13.4 SIDE CONSTRAINTS

In accordance with Section 11.5, the side constraints are here defined as

$$30 \text{ in} \leqslant L \leqslant 50 \text{ in},$$

8 in $\leqslant C \leqslant$ 20 in,

0.02 in $\leqslant t_s \leqslant$ 0.1 in,

0.05 in $\leqslant t_i \leqslant$ 0.2 in   for all i.

13.5   METHOD OF SOLUTION

We attempt the problem by using two different unconstrained minimization algorithms involving penalty functions:

(i)

$$\text{Minimize} f(X, r) = M(X) - r \sum_{j=1}^{3} \frac{1}{g_j(X)}. \tag{13.7}$$

The position of the spars is fixed at (a) 45 per cent chord for two cells; (b) 20 per cent and 70 per cent chord for three cells; and (c) 20 per cent, 45 per cent, and 70 per cent chord for four cells.

(ii)

$$\text{Minimize} \phi(X, X_p) = \frac{1}{M(X_p) - M(X)} - \frac{1}{R} \sum_{j=1}^{3} \frac{1}{g_j(X)}. \tag{13.8}$$

Algorithm (i) has been used in the problems in Chapters 11 and 12, and algorithm (ii) is similar to the method proposed by Fiacco and McCormick[7], except that a large positive arbitrary constant R, taken as unity, has been here introduced to accelerate the convergence of $\phi(X, X_p)$ towards the optimal value of $M(X)$.

For frequency calculations, the partial integro-differential equations arrived at by Krishnamurthy and Joga Rao[8,9] have been suitably used for a general multicell tube of arbitrary cross-section. Also, the secondary effects of shear lag, transverse shear, and longitudinal inertia on flexural vibration of symmetric tubes (see Katarya[10]) are considered.

13.6   RESULTS AND DISCUSSION

The results for a cantilevered rectangular wing with NACA 0010 airfoil cross-section have been obtained by considering four different problems with cells varying from 1 to 4. This has been done since the number of cells N has to be an integer, though the other design variables can be real numbers leading to the mixed-mode problem. The material constants are $E = 10^7$ psi, $\rho = 0.098$ lb/in^3, and $\nu = 0.3$.

Figure 13.1 depicts the complete optimization procedure for the cantilevered rectangular wing. Using Eq. (13.7), we get the results of wing optimization under a uniform load of 1000 lb, as given in Table 13.1. The table also shows the initial and final design vectors along with the number of minimizations required. It should be noted that a two-cell structure gives the minimum weight. Table 13.2 lists the results obtained for a tip moment of 20,100 lb in. Here again, the two-cell structure provides the minimum weight. In both cases, the behaviour constraints, stress, and frequency are satisfied at the optimal value and an increasing number of side constraints becomes active near the optimal values as the number of cells multiplies.

By using the ϕ-function [see Eq. (13.8)], optimization is achieved for the two-cell structure under uniform loading with one initial design vector and two values of constant R. The results obtained by using Eqs. (13.7) and (13.8) with the same initial vector are given in Table 13.3.

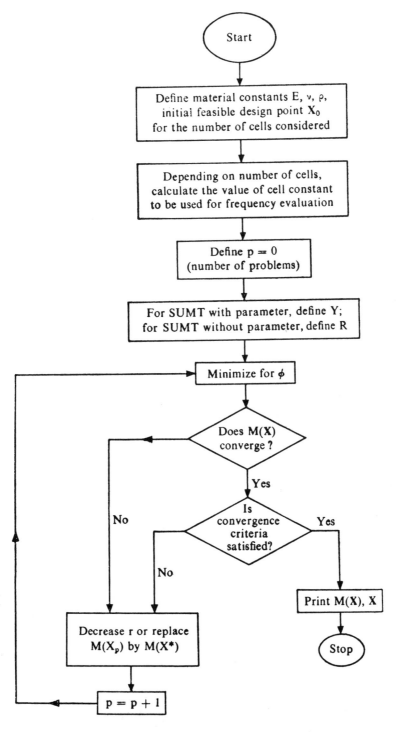

Fig. 13.1   Flowchart for Optimization of Cantilevered Rectangular Wing.

Table 13.1 Wing Optimization under Uniform Loading (without Parameter) (total load 1000 lb)

Number of cells	Number of unconstrained minimizations		L (in)	C (in)	t_s (in)	$t_{spar\,1}$ (in)	$t_{spar\,2}$ (in)	$t_{spar\,3}$ (in)	$f(X)$	ω (cps)	σ_s (psi)	$M(X)$ (in³)
1	5	Initial	42	14.50	0.1000				154.616	31.86		123.693
		Final	37.62	12.74	0.0770				75.150	34.96	29,983	75
2	5	Initial	33.42	16.40	0.0429	0.1998			72.428	32.80		57.942
		Final	30.40	16.87	0.0310	0.1836			41.047	34.98	29,987	41.024*
3	5	Initial	33.39	16.03	0.0450	0.0560	0.1780		72.010	34.02		57.608
		Final	30	16.25	0.0330	0.1355	0.1538		43.630	34.96	29,989	43.580
4	5	Initial	32	15	0.0470	0.1100	0.1100	0.1100	73.757	33.67		59.005
		Final	30	15.05	0.0405	0.0500	0.1600	0.0500	46.218	34.89	29,978	46.201

*Optimal solution

Table 13.2 Wing Optimization under Constant Tip Moment (without Parameter) (tip moment 20,100 lb in)

Number of cells	Number of unconstrained minimizations		L (in)	C (in)	t_s (in)	$t_{spar\,1}$ (in)	$t_{spar\,2}$ (in)	$t_{spar\,3}$ (in)	$f(X)$	ω (cps)	σ_s (psi)	$M(X)$ (in³)
1	6	Initial	40	14	0.1000				142.175	33.86		113.740
		Final	38.32	13.25	0.0690				71.239	34.99	29,976	71.217
2	6	Initial	40	14	0.0600	0.1000			110.179	28.64		73.453
		Final	31.28	17.80	0.0335	0.1959			48.070	34.98	30,000	48.057*
3	6	Initial	36	16	0.0550	0.0501	0.1802		91.806	31.99		73.445
		Final	33.34	17.66	0.0364	0.0500	0.1641		52.305	34.97	29,994	52.275
4	6	Initial	36	15.20	0.0600	0.1100	0.1100	0.1100	102.143	30.48		81.714
		Final	35.08	19.32	0.0293	0.0500	0.0879	0.0500	50.461	34.97	29,985	50.436

*Optimal solution

Table 13.3 Comparison between Interior Penalty Functions (with and without Parameters) in Optimization (uniform loading 1000 lb)

Type of penalty function	Number of unconstrained minimizations		L (in)	C (in)	t_s (in)	$t_{spar\ 1}$	$f(X)$	$\phi(X)$	ω (cps)	σ_s (psi)	R	$M(X)$ (in^3)
		Initial	33.20	13.64	0.0594	0.2000	78.823		32.50			63.058
With parameter	5	Final	30.87	13.28	0.0534	0.1999	52.102		34.89	29,940		52.050
Without parameter	4	Final	31.10	13.30	0.0536	0.2000		6.253	34.71	29,963	1	52.655
Without parameter	2	Final	31.39	12.92	0.0585	0.2000		1.823	34.83	29,981	26.72	54.823

For the symmetric airfoil wing problem, different starting points and convergence criteria give varying optimal values for some minimizations. From Table 13.3, we find that, for higher values of R, the function minimization by the penalty function approach without parameters proves to be faster than the one with parameters. From a computation point of view, the penalty function with parameters poses fewer problems.

It may be concluded from Figs. 13.2 and 13.3 that raising the number of cells beyond two does not lead to any substantial reduction or increase in weight. Figures 13.4 and 13.5 indicate the trends of function behaviour in reaching the minimum weight of the wing.

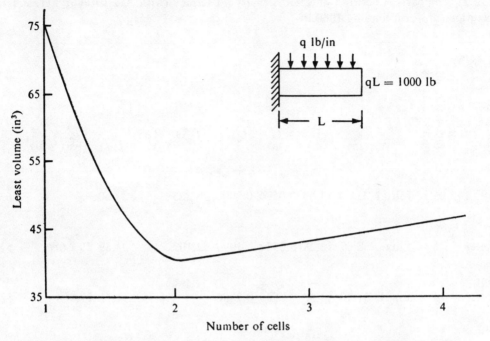

Fig. 13.2   Least Volume as Function of Number of Cells for Uniform Loading.

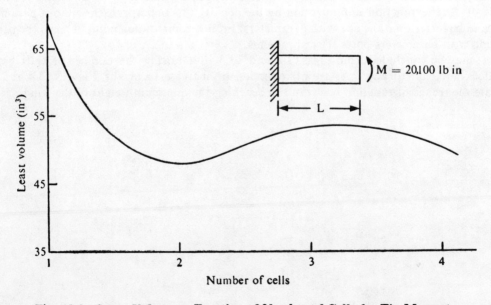

Fig. 13.3   Least Volume as Function of Number of Cells for Tip Moment.

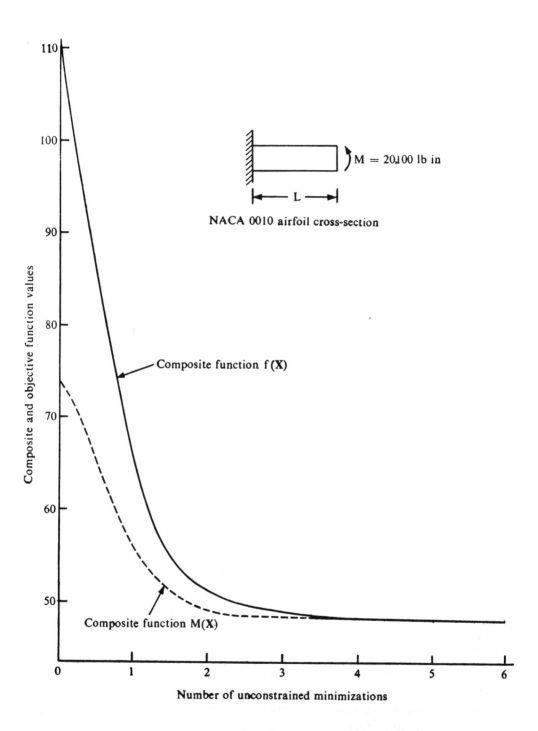

Fig. 13.4   Function Behaviour in Reaching Minimum for Two-Cell Structure with Tip Moment.

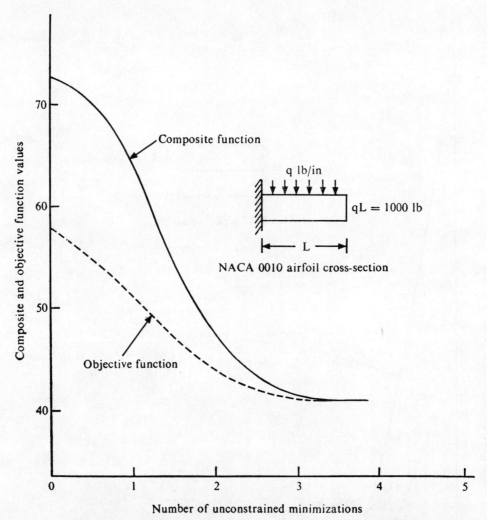

**Fig. 13.5   Function Behaviour in Reaching Minimum for
Two-Cell Structure with Uniform Loading.**

REFERENCES

1   Rubin, C. P., Minimum weight design of complex structures subject to a frequency constraint, *AIAA J.*, 8, 5, 923–27, 1970.

2   Fox, R. L. and Kapoor, M. P., Structural optimization in the dynamic response regime: a computational approach, *AIAA J.*, 8, 10, 1798–804, 1970.

3   Box, M. J., A new method of constrained optimization and a comparison with other methods, *Computer J.*, 8, 2, 42–52, 1965.

4   Fiacco, A. V. and McCormick, G. P., The sequential unconstrained minimization technique for nonlinear programming: a primal dual method, *Management Sc.*, **10**, 2, 360–66, 1964.

5   Fiacco, A. V. and McCormick, G. P., Computational algorithm for the sequential un-constrained minimization technique for nonlinear programming, *Management Sc.*, **10**, 4, 601–17, 1964.

6   Zoutendijk, G., Nonlinear programming: a numerical survey, *SIAM J. Cont.*, **4**, 2, 194–210, 1966.

7   Fiacco, A. V. and McCormick, G. P., The sequential unconstrained minimization technique (SUMT) without parameters, *Opns. Res.*, **15**, 5, 820–27, 1967.

8   Krishnamurthy, A. V. and Joga Rao, C. V., General theory of vibration of cylindrical tubes, Part I—formulation, *J. Aero. Soc. Ind.*, **20**, 1, 65–84, 1968.

9   Krishnamurthy, A. V. and Joga Rao, C. V., General theory of vibration of cylindrical tubes, Part III—uncoupled flexural vibration of closed tubes, *J. Aero. Soc. Ind.*, **20**, 3, 157–74, 1968.

10 Katarya, R., Optimization of multi-cellular wings under strength and vibrational con-straints for simple loading, M. Tech. Thesis, Department of Aeronautical Engineering, IIT, Kanpur, 1973.

14
Minimum Cost Design of Grid Floor

14.1 INTRODUCTION

The structural problems discussed so far involved only one material, and the optimization has been carried out to minimize the structural weight. The mathematical techniques dealt with in Chapters 7–9 can be applied also to structures made up of two or more materials such as composite and reinforced concrete structures. In this chapter, we shall discuss the optimal design problem of a reinforced concrete grid floor.

In civil engineering structures, the minimum weight does not necessarily mean that the cost too is a minimum. Various costs, such as that of shuttering, which does not form a part of the structure, have also to be taken into account in arriving at the minimum cost. We shall now analyze the problem of the minimum cost design of a reinforced concrete floor with orthogonal grid. The ultimate strength design concept postulated by Ferguson[1] which has proved to be logical and economical for reinforced concrete structures is applied in our analysis. (For details on the developments that have taken place in the optimization of reinforced concrete structures, see Palejs[2], Mast[3], Kohli[4], Freyermuth[5], Derecho[6], and Aperghis[7].)

14.2 FORMULATION

A grid floor with given dimensions and a specified load is designed to obtain the minimum

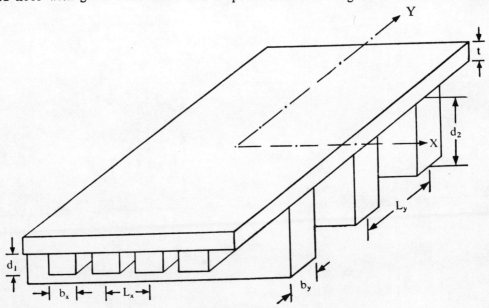

Fig. 14.1 Typical Section of Grid Floor.

cost. The grids are analyzed by applying the elastic theory which introduces suitable load factors for the dead and live loads. The sections of each grid are designed by employing the ultimate strength theory. Figure 14.1 shows the typical section of a grid floor.

14.3 OBJECTIVE FUNCTION

The objective function for our problem is the cost per unit area of the floor, which is given by

$$f(X) = (L_x L_y t + b_x d_1 L_y + b_y d_2 L_x - b_x b_y d_1)$$
$$+ CR[(A_{stsx} + A_{stbx})L_x + (A_{stsy} + A_{stby})L_y], \tag{14.1}$$

where CR is the cost ratio of the unit cost of steel to concrete, d_1 and d_2 are the depth of rib below the concrete soffit of the slab along the X- and Y-direction, b_x and b_y are the width of rib along the X- and Y-direction, P_{stsx} and P_{stsy} are the percentage of steel in the slab, and P_{stbx} and P_{stby} are the percentage of steel in the beam.

14.4 BEHAVIOUR CONSTRAINTS

The behaviour constraints are based on the strength requirements of the slab and the beam to resist the maximum bending moment along the X- and Y-direction. This implies that

(i) the ultimate moment of resistance of the slab in the X- and Y-direction is greater than or equal to the maximum moment of resistance due to the applied load along the X- and Y-direction, i.e.,

$$M_{rsx}^U \geqslant M_{\max sx}^U,$$
$$M_{rsy}^U \geqslant M_{\max sy}^U; \tag{14.2}$$

(ii) the ultimate moment of resistance of the beam in the X- and Y-direction is greater than or equal to the maximum moment due to the applied loads along the X- and Y-direction, respectively, i.e.,

$$M_{rbx}^U \geqslant M_{\max bx}^U,$$
$$M_{rby}^U \geqslant M_{\max by}^U; \tag{14.3}$$

(iii) sufficient ductility is provided, and, as per Indian Standards, the depths of the compressive stress block a both in the slab and the beam do not exceed 0.43 times their respective effective depth, i.e.,

$$a_{sx} \geqslant 0.43t,$$
$$a_{sy} \geqslant 0.43t,$$
$$a_{bx} \geqslant 0.43d, \tag{14.4}$$
$$a_{by} \geqslant 0.43d.$$

14.5 SIDE CONSTRAINTS

The restrictions imposed on the thickness of the slab, and the width of the beams and their spacings in the X- and Y-direction are

$$t \geqslant 10 \text{ cm}, \quad b_y \geqslant 15 \text{ cm}, \quad b_x \geqslant 15 \text{ cm}, \quad L_x \leqslant 250 \text{ cm}, \quad L_y \leqslant 250 \text{ cm}. \tag{14.5}$$

14.6 METHOD OF SOLUTION

The grid floor is idealized as an orthotropic plate of constant thickness for estimating stresses and deformations due to flexure. The basic equations involved in the analysis are discussed in

Section 3.5. For the sake of continuity, we repeat some of the pertinent equations:

$$M_x = -D_x\left(\frac{\partial^2 w}{\partial x^2} + v_y \frac{\partial^2 w}{\partial y^2}\right),$$

$$M_y = -D_y\left(\frac{\partial^2 w}{\partial y^2} + v_x \frac{\partial^2 w}{\partial x^2}\right), \tag{14.6}$$

$$M_{xy} = -2D_{xy}\frac{\partial^2 w}{\partial x\, \partial y},$$

where

$$D_x = \frac{E_x t^3}{12(1 - v_x v_y)},$$

$$D_y = \frac{E_y t^3}{12(1 - v_x v_y)}, \tag{14.7}$$

$$D_{xy} = G_{xy}\frac{t^3}{12}.$$

The governing equation for flexure of an orthotropic plate may be written as

$$D_x\frac{\partial^4 w}{\partial x^4} + (D_x v_y + D_y v_x + 4D_{xy})\frac{\partial^4 w}{\partial x^2 \partial y^2} + D_y \frac{\partial^4 w}{\partial y^4} = \rho(x, y). \tag{14.8}$$

The equivalent orthotropic plate of the grid has two bending stiffnesses along the two coordinate directions X and Y and are governed respectively by the second moment of area I_x and I_y per unit width of the plate. The material properties E and v are considered to be the same along the two directions. Hence, the values of D_x, D_y, and D_{xy} in Eq. (14.8) are

$$D_x = \frac{EI_x}{b},$$

$$D_y = \frac{EI_y}{b}, \tag{14.9}$$

$$D_{xy} = \frac{G_{xy}}{3}\left(\frac{b_x d_1^3}{a} + \frac{b_y d_2^3}{b}\right).$$

The solution for $w(x, y)$, which satisfies the boundary conditions, is chosen in the form of a series of undetermined coefficients. These are determined by substituting the solution in (14.8). By knowing $w(x, y)$, we can obtain the maximum bending and twisting moments for the plate from Eqs. (14.6).

For optimization, the constrained programming problem is transformed into an unconstrained programming problem by augmenting the objective function with the constraints. Using a generalization of the Davidon-Fletcher-Powell method suggested by Stewart[8] to solve unconstrained optimization problems without the derivatives, we have obtained the optimal solution. The subroutine proposed by Fletcher[9] has been incorporated into the main program for optimization (see Narayanan[10]).

14.7   RESULTS AND DISCUSSION

The results obtained for a typical grid floor are given in Table 14.1. From the data listed, it is evident that when relative cost ratios are included in the formulation of an optimum design problem, the resulting design becomes economical. The optimal design is a 1.2 metre square grid as against the normal 1 metre square grid. This indirectly saves the labour involved in the construction and designing of form work.

Table 14.1   Optimum Design of Grid Floor

Given

$$L = B = 18.83 \text{ m},$$

cubic strength of concrete at end of 28 days $= 200 \text{ kg/cm}^2$,

yield stress in steel $= 2400 \text{ kg/cm}^2$ (for bars > 20 mm dia)

$\qquad\qquad\qquad\quad = 2000 \text{ kg/cm}^2$ (for bars $\leqslant 20$ mm dia),

live load $= 75 \text{ kg/m}^2$.

Initial design	Final design
$t = 20$ cm	$t = 18.87$ cm
$L_x = 100$ cm	$L_x = 117.33$ cm
$L_y = 100$ cm	$L_y = 117.33$ cm
$d_1 = 60$ cm	$d_1 = 22.35$ cm
$d_2 = 60$ cm	$d_2 = 22.35$ cm
$b_x = 30$ cm	$b_x = 15$ cm
$b_y = 30$ cm	$b_y = 15$ cm
$P_{stmb}/P_{stbx} = 0.1$	$P_{stmb}/P_{stbx} = 0.3567$
$P_{stmb}/P_{stby} = 0.1$	$P_{stmb}/P_{stby} = 0.3567$
$P_{stms}/P_{stsx} = 0.1$	$P_{stms}/P_{stsx} = 0.5670$
$P_{stms}/P_{stsy} = 0.1$	$P_{stms}/P_{stsy} = 0.5670$
	Relative cost $= 58.76$

REFERENCES

1   Ferguson, P. M., Reinforced Concrete Fundamentals, 3rd edn., Wiley, New York, 1973.

2   Palejs, A. A., Command language interface for automatic design of structures in three dimensions, Am. Conc. Inst. Fall Convention, St. Louis, 1970.

3   Mast, P. E., Computer Programs for Analysis and Design of Flat Plates and Continuous Concrete Frames, SR 017.01D, Portland Cement Assoc., Skokie, Illinois, 1968.

4   Kohli, J. P., Optimum design of concrete spread footing by computer, ACI J. Proc., 65, 5, 384–89, 1968.

5   Freyermuth, C. L., Computer Program for Analysis and Design of Simple Span Precast Prestrained Highway Railway Bridge, SR 033.10E, Portland Cement Assoc., Skokie, Illinois, 1968.

6   Derecho, A. J., Computer Program for the Analysis and Design of Concrete Wall-Beam Frames, Special Report, Portland Cement Assoc., Skokie, Illinois, 1969.

7   Aperghis, G. G., Slab bridge design and drawing—an automated process, *Proc. Inst. Civil Engrs.*, **46**, 1, 55–75, 1970.

8   Stewart, III, G. W., A modification of Davidon's minimization method to accept difference approximations to derivatives, *J. ACM*, **14**, 1, 72–83, 1967.

9   Fletcher, R., Subroutines for Minimization by Quasi-Newton Method, R 7125, AERE, Harwell, 1971.

10   Narayanan, G. V., Minimum cost design of the beam and grid floors, M. Tech. Thesis, Department of Civil Engineering, IIT, Kanpur, 1974.

15

Nonlinear Free Vibration of Beams

15.1 INTRODUCTION

Burgreen[1] and Sato and Sato[2] studied the problem of nonlinear vibration of beams having structural nonlinearity. (For additional details on this problem, see Eisley[3].) The nonlinearity of beams is due to large amplitude. The additional inplane stresses caused by large amplitude stiffen the structure, thus raising the frequency response. Sato and Sato[2] studied the free vibration of a beam with a central mass having structural nonlinearity. They assumed a linear mode shape satisfying the boundary conditions and observed that the frequency increases with the corresponding amplitude which in turn rises with the increase in mass. In this chapter, the problem of free vibration of two types of simply-supported beams with material nonlinearity—one with central mass and the other without—is studied. Iyengar[4] has shown that the frequency response of these two types of beams can also be obtained by employing the finite element and perturbation methods, and that the frequency decreases with increase in amplitude. He has further demonstrated that when using the perturbation method, the evaluation of higher order terms is too complex, whereas when using the finite element technique, the iteration can be started only when the frequency and mode shape are known.

In this chapter, we shall establish that the programming technique (unconstrained minimization) can be successfully employed for studying the nonlinear vibration problems. It may further be noted that both the mode shape and frequency can also be progressively refined by using other mathematical programming techniques.

15.2 MATERIAL STRESS-STRAIN LAW

Very few materials are linear over a wide range of strains and ideally plastic in the inelastic range. Except in the region near the origin of the stress-strain relation, most materials are nonlinearly elastic. The stress-strain relation for the material is assumed (see Iyengar[4]) to be of the form

$$\sigma_x = A'\epsilon_x - B'\epsilon_x^m, \tag{15.1}$$

where A', B', and m are constants dependent on the material and temperature.

15.3 FORMULATION

For a slender beam with a rectangular cross-section, we formulate the problem in terms of the Lagrangian L which involves the potential and kinetic energies of the system. When minimized, the system gives the mode shapes and its corresponding frequencies. The Lagrangian L may be written (see Iyengar[4]) as

$$L = \frac{A'I}{2} \int_0^{L_b} \int_0^{2\pi/\omega} (\partial^2 v/\partial z^2)^2 \, dz \, dt - \frac{B'(h/2)^{m+2}b}{m+2} \int_0^{L_b} \int_0^{2\pi/\omega} (\partial^2 v/\partial z^2)^{m+1} \, dz \, dt$$

$$- \frac{m_b}{2} \int_0^{L_b} \int_0^{2\pi/\omega} (\partial v/\partial t)^2 \, dz \, dt - M_e \int_0^{2\pi/\omega} (\partial v/\partial t)^2 v \Big|_{z=1/2} \, dt, \tag{15.2}$$

where m_b, L_b, b, and h are respectively mass, length, breadth, and thickness of the beam, v is the transverse displacement, and M_e the external mass on the beam. For vibration studies which do not involve external mass, the last term of Eq. (15.2) is dropped.

The minimization of (15.2) becomes rather difficult for any general value of m, and to illustrate the minimization technique, let us assume m to be 5. Equation (15.2) may thus be written as

$$L = \frac{A'I}{2} \int_0^{L_b} \int_0^{2\pi/\omega} (\partial^2 v/\partial z^2)^2 \, dz \, dt - \frac{B'(h/2)^7 b}{7} \int_0^{L_b} \int_0^{2\pi/\omega} (\partial^2 v/\partial z^2)^6 \, dz \, dt$$
$$- \frac{m_b}{2} \int_0^{L_b} \int_0^{2\pi/\omega} (\partial v/\partial t)^2 \, dz \, dt - M_e \int_0^{2\pi/\omega} (\partial v/\partial t)^2 v \bigg|_{z=1/2} dt. \tag{15.3}$$

15.4 EQUIVALENT PROGRAMMING PROBLEM

In our study, the unconstrained minimization technique is employed to obtain the frequency ω and mode shape. The design variables represent the displacements at discrete points of the beam and ω. Since we consider the frequency a variable, the method gains an advantage over the classical Rayleigh-Ritz and Galerkin techniques, where ω is treated as a constant.

Programming techniques have been well developed for discrete variables; hence, in Eq. (15.3), the derivatives have to be replaced by their finite differences and the integral sign by a summation sign. The problem becomes quite intricate if $v(z, t)$ is assumed to be the unknown function of z and t. In this case, the number of variables gets doubled without perhaps a corresponding increase in the information obtained. The time function should therefore be so assumed that it satisfies the initial conditions. Since ω is still a variable, the frequency can be calculated as accurately as desired within the limitations implied in the foregoing assumptions. We assume $v(z, t)$ to be of the form

$$v(z, t) = \phi(z) \cos \tau \qquad (\tau = \omega t). \tag{15.4}$$

By substituting the expressions for $v(z, t)$ and t from Eq. (15.4) in (15.3), L can be rewritten as

$$L = \frac{A'I}{2} \int_0^{L_b} \int_0^{2\pi} (\frac{d^2\phi}{dz^2})^2 \cos^2 \tau \, dz \, d\tau - \frac{5B'(h/2)^7 b}{56} \int_0^{L_b} \int_0^{2\pi} (\frac{d^2\phi}{dz^2})^6 \, dz \cos^6 \tau \, d\tau$$
$$- \frac{m_b}{2}\omega^2 \int_0^{L_b} \int_0^{2\pi} \sin^2 \tau \, \phi^2(z) \, dz \, d\tau - M_e\omega^2\phi^2(z) \bigg|_{z=L_b/2} \tag{15.5}$$

The boundary conditions are

$$\phi(0) = \phi(L_b) = 0,$$
$$\frac{d^2\phi}{dz^2}(0) = \frac{d^2\phi}{dz^2}(L_b) = 0. \tag{15.6}$$

In terms of the nondimensional parameter ξ $(=z/L_b)$, we put

$$\Delta z = L_b \cdot \Delta \xi \tag{15.7a}$$

and, since $L_b = n\Delta z$,

$$\Delta \xi = 1/n. \tag{15.7b}$$

The beam is divided into a number of segments, with $\phi_0, \phi_1, \ldots$ denoting the displacements at the end of each segment. Then, by substituting Eqs. (15.7) in Eqs. (15.5) and (15.6) and replacing the derivatives by the equivalent differences, we may express (15.5) in the form

$$L = \sum_{i=1}^{n-1} \left[\frac{A'I}{2} \left\{ \left(\frac{\Delta^2 \phi}{\Delta \xi^2} \right)_i \right\}^2 \frac{1}{L_b^4} - \frac{5B'b(h/2)^7}{56} \left\{ \left(\frac{\Delta^2 \phi}{\Delta \xi^2} \right)_i \right\}^6 \frac{1}{L_b^{12}} - \frac{m_b}{2} \omega^2 (\phi_i)^2 \right] \frac{L_b}{n} - M_e \omega^2 \phi_k^2 \Big|_{k=(n-1)/2},$$

(15.8)

where

$$\left(\frac{\Delta^2 \phi}{\Delta \xi^2} \right)_i = n^2 (\phi_{i-1} - 2\phi_i + \phi_{i+1}).$$

The boundary conditions (15.6) may be reduced to

$$\phi_1 = 0, \qquad \phi_0 = -\phi_2, \qquad \phi_n = 0, \qquad \phi_{n-1} = -\phi_{n+1}.$$

(15.9)

It is required that L be minimized subject to conditions (15.9) with the variables $\phi_0, \phi_1, \ldots$ being positive definite.

15.5 MINIMIZATION TECHNIQUE

The problem as defined by Eqs. (15.8) and (15.9) is a constrained optimization problem. It is transformed into a sequence of unconstrained minimization problems by constructing the P-function as

$$P_f = L + r \Sigma \frac{1}{g_i(\phi_i)}.$$

(15.10)

In order to commence the minimization algorithm, we must know the feasible point that satisfies the constraints. In our study, the feasible point is taken to be that derived from linear analysis.

When the minimization is carried out from three feasible points, it is observed that the minimum value of the objective function differs from one initial point to the other. This means that the minimum obtained is not global but local. The minima are derived from the minimum value of the objective function as the ordinate and the frequency as the abscissa. The frequency corresponding to the minimum of these minima is chosen as the optimum value.

15.6 RESULTS AND DISCUSSION

Tables 15.1–15.4 show the results obtained for a uniform beam that has no central mass for different values of A' and B' along with those obtained from the finite element and perturbation methods. It may be noted that, for the first mode, the results of all the three methods—perturbation, finite element, and programming—are quite similar, whereas for the second mode, the results of only the finite element and programming methods are close to each other.

Tables 15.5–15.8 give the results for a beam with a central mass for different values of A' and B'. As can be seen, the frequency decreases with the increase in amplitude and this effect increases with the addition in the value of the external mass M_e.

The time required for minimizing the P-function for one set of starting points increases with the number of segments into which the beam is divided. The average time is about 70 seconds on the IBM 7044 computer for 12 segments of the beam.

We have therefore established that the unconstrained minimization technique can be applied in studying nonlinear vibration problems. Similarly, the other mathematical programming techniques can be successfully used to obtain the response of nonlinear vibration problems.

Table 15.1 Comparison with Finite Element and Perturbation Methods (first mode)

Given

$$A' = 45.6 \times 10^2, \qquad B' = 2.025 \times 10^{16}, \qquad m = 5.$$

Nondimensional amplitude	Perturbation method	Finite element method (18 divisions)	Programming method (14 divisions)
4	122.07	121.9398	121.8999
6	121.98	121.8699	121
8	121.67	120.8794	120.1946
10	120.61	117.0184	118.0294

Table 15.2 Comparison with Finite Element and Perturbation Methods (first mode)

Given

$$A' = 43.207 \times 10^2, \qquad B' = 2.035 \times 10^{16}, \qquad m = 5.$$

Nondimensional amplitude	Perturbation method	Finite element method (18 divisions)	Programming method (14 divisions)
4	118.8801	118.4235	118.9992
6	118.7723	118.4004	118.8949
8	118.4816	117.7815	116.9549
10	117.8664	115.3043	115.19997

Table 15.3 Comparison with Finite Element and Perturbation Methods (second mode)

Given

$$A' = 45.6 \times 10^2, \qquad B' = 2.025 \times 10^{16}, \qquad m = 5.$$

Nondimensional amplitude	Perturbation method	Finite element method (20 divisions)	Programming method (16 divisions)
2	486.9725	483.3395	482.6622
2.2	486.2068	479.9330	480.9478
2.4	485.1993	471.6159	474.9992
2.6	483.9028	463.2658	461.4970

Table 15.4   Comparison with Finite Element and Perturbation Methods (second mode)

Given

$$A' = 42.399 \times 10^2, \qquad B' = 2.3385 \times 10^{16}, \qquad m = 5.$$

Nondimensional amplitude	Perturbation method	Finite element method (20 divisions)	Programming method (16 divisions)
2	469.4381	466.6505	466.4599
2.2	468.6383	462.2200	464.2999
2.4	467.5861	455.2748	460.4998
2.6	466.2315	446.6111	455.9996

Table 15.5   Comparison with Finite Element and Perturbation Methods (first mode)

Given

$$A' = 45.6 \times 10^2, \qquad B' = 2.025 \times 10^{16}, \qquad m = 5,$$
$$M_e/m_b = 0.25.$$

Nondimensional amplitude	Perturbation method	Finite element method (18 divisions)	Programming method (12 divisions)
4	99.6395	99.4825	99.5528
6	99.5399	99.4769	99.4523
8	99.2713	99.3873	99.1057
10	98.7024	98.7150	98.6000

Table 15.6   Comparison with Finite Element and Perturbation Methods (first mode)

Given

$$A' = 43.207 \times 10^2, \qquad B' = 2.035 \times 10^{16}, \qquad m = 5,$$
$$M_e/m_b = 0.25.$$

Nondimensional amplitude	Perturbation method	Finite element method (18 divisions)	Programming method (12 divisions)
4	96.9886	96.6129	96.0053
6	96.8858	96.6118	95.8056
8	96.6083	96.5789	95.5058
10	96.0206	96.1681	95.1103

Table 15.7   Comparison with Finite Element and Perturbation Methods (first mode)

Given

$$A' = 45.6 \times 10^2, \qquad B' = 2.025 \times 10^{16}, \qquad m = 5,$$
$$M_e/m_b = 0.5.$$

Nondimensional amplitude	Perturbation method	Finite element method (18 divisions)	Programming method (12 divisions)
4	86.2451	86.0663	86.3059
6	86.1569	86.0661	86.1059
8	85.9188	86.0634	86.0065
10	85.0163	86.0418	85.5220

Table 15.8   Comparison with Finite Element and Perturbation Methods (first mode)

Given

$$A' = 43.207 \times 10^2, \qquad B' = 2.035 \times 10^{16}, \qquad m = 5,$$
$$M_e/m_b = 0.5.$$

Nondimensional amplitude	Perturbation method	Finite element method (18 divisions)	Programming method (12 divisions)
4	83.9506	83.5836	83.6102
6	83.8594	83.5836	83.3368
8	83.6135	83.5831	83.1375
10	83.0927	83.5761	83.0061

REFERENCES

1   Burgreen, D., Free vibrations of a pin ended column with constant distance between pin ends, *J. Appl. Mech.*, **18**, 2, 135–39, 1951.

2   Sato, K. and Sato, T., Free vibrations of a beam with a central mass, The Technology Report of the Tohoku University, Japan, Vol. 33, 149–54, 1968.

3   Eisley, J. G., Nonlinear deformation of elastic beams, rings and strings, *Appl. Mech. Rev.*, **16**, 9, 677–80, 1963.

4   Iyengar, N. G. R., Nonlinear high temperature effects on static and dynamic response of structural elements, Ph.D. Thesis, Department of Aeronautical Engineering, IIT, Kanpur, 1971.

16

Optimization of Wing Structures

16.1 INTRODUCTION

In our examples thus far, the constraints on optimization have been very few. However, in a real life problem, we encounter multiple constraints. Schmit and Thornton[1] considered a highly idealized double wedge airfoil with the total propulsive work as the objective function and restrictions on the root angle of attack, tip deflection, root stresses, and bending-torsion flutter Mach number. Stroud *et al*[2] obtained the minimum weight by idealizing the wing as an isotropic sandwich plate with constraints on strength, stability, and flutter. For Rao[3], strength, stability, frequency, and flutter were the major requirements in the optimization of wing structures. Giles[4] developed a procedure to include the interaction of external shape, aerodynamic loads, structural geometry, and fuel mass for a fully stressed design of aerodynamic surfaces.

In this study, we shall demonstrate the feasibility of introducing multiple behaviour constraints in the optimum design of a wing structure with design variables including aerodynamic parameters, such as sweep back angle, aspect ratio, and thickness to chord ratio, in addition to the usual gauge parameters. Before optimization is carried out, we shall conduct a parametric study of the behaviour quantities, e.g., maximum deflection, stresses, buckling load, and natural frequencies, to understand how these are influenced by the design variables.

For static and dynamic analysis, the aircraft wing is idealized by finite elements through the use of constant stress triangular membrane elements and rectangular shear panels for the skins and the webs, respectively. The stringers are represented by axial force members. Elastic buckling constraints are introduced by treating a typical portion of the wing skin as an isotropic stiffened plate.

16.2 FORMULATION

We deal with the minimum weight design of an idealized aircraft wing which is composed of several finite elements. The following wing design criteria are to be satisfied:

    (i) The elastic deflection at the tip of the wing should not exceed a prescribed limit.

    (ii) The stresses induced at the root and at the tip of the wing should not be more than the yield stress of the material. (This constraint is important since the skin thickness decreases from the root to the tip.)

    (iii) The stresses induced in the skin member should not cause any local buckling.

    (iv) The first two natural frequencies of the structure should be within the specified bounds.

Constraints (ii) and (iii) have to be satisfied at every point in the structure. This is a rather time-consuming procedure, but it can be simplified by checking the constraints at any one critical section rather than at every point.

The design variables include sweep back angle, aspect ratio, and thickness to chord ratio in order to assess the feasibility of optimization in their presence and to study their interaction with the external shape and mass distribution of the wing. To start with, there are 13 design variables representing the linearly decreasing mass distribution of the cover plate elements and web and flange elements of spars and ribs. These variables are reduced to nine since a preliminary parametric study shows that the linearly decreasing mass distribution of spars and ribs can be replaced by uniform mass distribution without any appreciable change in the behaviour constraints.

The optimization problem may be expressed in the form:

$$\text{Minimize } f(X) = \sum_{j=1}^{N} A_j \rho_j x_j + \sum_{j=1}^{R} L_j \rho_j x_j, \tag{16.1}$$

where

$f(X)$ = weight of the structure,

A_j = area of the j-th plate element,

N = number of plate elements,

ρ_j = density,

L_j = length of the j-th pin-jointed bar,

x_j = j-th design variable,

subject to

$$\delta_{\text{tip}}^{u} - \delta_{\text{tip}} \geqslant 0, \tag{16.2}$$

$$\sigma_{\text{root}}^{u} - \sigma_{\text{root}} \geqslant 0, \tag{16.3}$$

$$\sigma_{\text{tip}}^{u} - \sigma_{\text{tip}} \geqslant 0, \tag{16.4}$$

$$\sigma_{\text{b root}}^{u} - \sigma_{\text{b root}} \geqslant 0, \tag{16.5}$$

$$\sigma_{\text{b tip}}^{u} - \sigma_{\text{b tip}} \geqslant 0, \tag{16.6}$$

$$\omega_j^{l} \leqslant \omega_j \leqslant \omega_j^{u} \quad (j = 1, 2), \tag{16.7}$$

$$X_j^{l} \leqslant X_j \leqslant X_j^{u} \quad (j = 1, \ldots, n), \tag{16.8}$$

where the superscripts u and l stand for upper and lower limits, ω represents the natural frequency, σ_b is the buckling stress, and σ is the static stress. The relationship between the behaviour quantities δ_{tip}, σ_{root}, σ_{tip}, $\sigma_{\text{b root}}$, $\sigma_{\text{b tip}}$, and ω_j cannot be expressed in the closed form.

16.3 ANALYSIS

The prediction of the static and dynamic behaviour of a wing structure is an extremely complex problem. The finite element method of structural analysis which we shall apply entails representing the wing structure as an assembly of triangular membrane cover elements, rectangular shear web elements, and pin-jointed flange elements. The available computer core (32k, IBM 7044) eliminates the possibility of using higher order elements. For a detailed discussion on the derivation of stiffness and mass matrices and also the static and dynamic analysis procedure, the reader should refer to Rao[5].

The buckling load constraint is obtained by taking the portion of the grid stiffened plate between the spars and the ribs as an orthotropic, rectangular, simply-supported plate for which a closed-form solution exists (see Section 3.5).

16.4 METHOD OF SOLUTION

The constrained optimization problem (16.1)–(16.8) is solved as a sequence of unconstrained minimization problems by employing the interior penalty function approach. The Davidon-Fletcher-Powell variable metric method is used for minimization. The cubic interpolation technique of one-dimensional search, which makes use of the gradient, is applied for finding the step length.

Most automated design problems take considerable time to satisfy the prescribed convergence criteria even when the optimum design has almost been achieved. This requirement is fulfilled by having coarse convergence criteria in the early stages of optimization and refined convergence criteria in the later stages. Further, a major part of the optimization time is spent in time-consuming repeat analyses. To overcome this, a rapid reanalysis is the key factor in economizing on computer time.

The behaviour variables are linearly approximated whenever the changes in the design variables are small. Let σ_b be the buckling stress corresponding to a design X and σ_b^* the buckling stress corresponding to a perturbed design X^*. By knowing σ_b and its derivative at X, the buckling stress σ_b^* can be approximated as

$$\sigma_b^* = \sigma_b + \sum_{k=1}^{N} \frac{\partial \sigma_b}{\partial x_k} x_k, \tag{16.9}$$

provided

$$\frac{\Delta x_k}{x_k} = \frac{x_k^* - x_k}{x_k} < \epsilon. \tag{16.10}$$

The value of ϵ is taken to be 0.1 in the early stages of optimization and reduced in consecutive steps until it is 0.02 in the final stages. It may be noted that this technique alone has resulted in 30 per cent reduction in the total computer time.

16.5 NUMERICAL EXAMPLE

As already observed, the dimensions of the wing depend on design variables, such as aspect ratio, thickness to chord ratio, and sweep back angle. The main dimensions of the wing are obtained as

$$b(\text{semispan}) = (AR \cdot SA \cdot 0.5)^{1/2},$$
$$c(\text{chord}) = b \tan \lambda + SA/b, \tag{16.11}$$
$$t(\text{thickness}) = (c - 2b \tan \lambda)TR/0.8,$$

where AR is the aspect ratio, SA the surface area, and TR the taper ratio. The leading and trailing edge spars are assumed to be at 10 per cent and 90 per cent of the chord, respectively.

The wing data employed in our numerical study is shown in Table 16.1. Whenever the

Table 16.1 Wing Material Data for Numerical Computation

	Material	Titanium
Material properties	Modulus of elasticity	16.4×10^6 psi (1.05×10^6 kg/cm^2)
	Poisson's ratio	0.3
	Density	0.16 lb/in (0.0045 kg/cm^3)
Details of wing	Planform area	3542 ft^2 (318.78 m^2)
	Payload	4,00,000 lb ($\approx$2,00,000 kg)

design is changed, the main dimensions of the wing have to be re-evaluated since they depend on the design variables. Figures 16.1 and 16.2 depict the wing structure idealization and planform nodes, respectively.

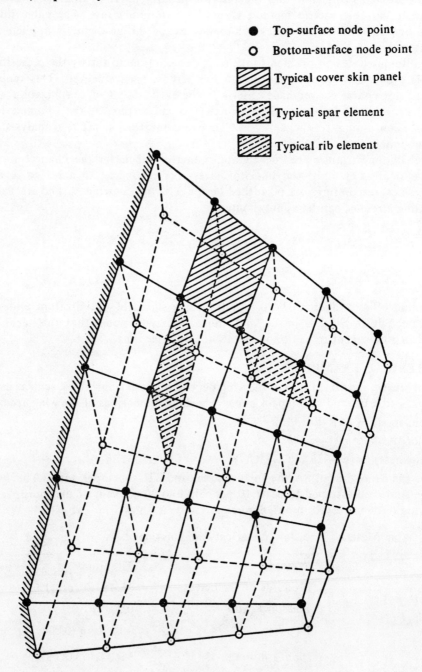

● Top-surface node point

○ Bottom-surface node point

Typical cover skin panel

Typical spar element

Typical rib element

Fig. 16.1 Wing Structure Idealization.

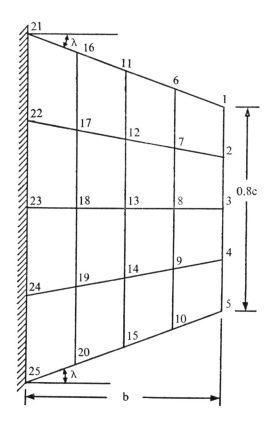

Fig. 16.2   Planform Nodes and Important Dimensions of Wing Structure.

The finite element idealization of the wing is carried out with 50 nodes and 172 elements, the latter comprising 64 triangular membrane elements, 36 rectangular shear panels, and 72 bar elements. For the eigenvalue problem, the number of degrees of freedom is reduced to 20 by using the plate flexural assumption and static condensation technique. Table 16.2 gives the results of the initial parametric study conducted for observing the influence of design variables on behaviour quantities. This study reveals that a linearly decreasing mass distribution of spars and ribs can be replaced by a uniform distribution without causing any appreciable change in the behaviour constraints, thus reducing the number of variables to nine.

The minimum weight design problem is initially solved by employing the linearly approximated reanalysis, the results of which are stated in Table 16.3. To assess the time saved in this analysis, the same problem is solved by employing the exact reanalysis in the third and fourth unconstrained minimizations. The results given in Table 16.4 indicate that the total computation time is reduced by about 30 per cent; however, the minimum weight gets increased by 1.5 per cent. In order to test the type of optimum obtained, that is, whether it is local or global in the design space, the same problem is rerun with a different starting point. By doing this, we observe that the new starting point provides approximately the same result as that arrived at by using the original starting point.

Table 16.2   Results of Initial Parametric Study

Percentage change in design variables (initial values)	Percentage change in behaviour quantities							
	$f(X)$ (49.5)	δ_{tip} (44.1)	σ_{root} (41.8)	σ_{tip} (9.4)	$\sigma_{b\,root}$ (41)	$\sigma_{b\,root}$ (41)	ω_1 (2.46)	ω_2 (3.95)
$x_1(0.05)$* +50	−11	+8.60	0	+49.	0	+53	+14.70	+8.10
−50	+11	−3.60	0	−24.10	0	−25.40	−9.35	−5.80
$x_2(0.35)$ +50	+55	−35	−33.20	−44.50	−33.30	−45.40	−6.90	−4.30
−50	−57	+145	+96	−490	+97	+520	+31	+13.20
$x_3(0.05)$ +50	−0.2	−0.23	0	+0.60	0	+0.31	+0.50	+0.17
−50	+0.2	+0.15	0	−0.32	0	−0.23	−0.15	−0.25
$x_4(0.35)$ +50	+1	−0.68	−0.16	−0.60	−0.16	−0.54	−0.27	−0.42
−50	−1	+2.60	0	+4.60	0	+1.65	−0.54	−0.25
$x_5(0.20)$ +50	−0.4	+0.46	0	+1	0	0	+0.25	−0.43
−50	+0.4	−0.30	0	−0.64	0	−0.46	−0.25	−0.51
$x_6(1.40)$ +50	+2	−1.36	−0.32	−1.20	−0.32	−1.10	−0.54	−0.84
−50	−2	+5.20	0	+9.20	+0	+3.30	−1.08	−0.51
$x_7(0.05)$ +50	−0.1	0	0	0	0	0	+0.15	+0.08
−50	+0.1	0	0	0	0	0	−0.15	−0.08
$x_8(0.35)$ +50	−0.5	−0.07	0	−0.04	0	0	−0.41	−0.3
−50	−0.5	+0.38	−0.08	+0.11	−0.08	−0.5	+0.41	+0.8
$x_9(0.20)$ +50	−0.1	0	0	0	0	0	+0.25	+0.17
−50	+0.1	0	0	0	0	0	−0.25	−0.17
$x_{10}(1.40)$ +50	1.0	−0.15	0	−0.07	0	0	−0.82	−0.70
−50	−1.0	+0.76	−0.16	+0.21	−0.16	−1	+0.82	+0.17
$x_{11}(1.50)$ +50	+1.0	+188	+60	+115	+60	115	−36.6	−13.40
−50	+0.6	−81	−65	−38.40	−65	−65	+115	+48.60
$x_{12}(0.005)$ +50	2.4	−55	−33.50	−33	−33.60	−33.40	+46.4	+46
−50	−2.4	+294	+101	+99	101.10	100	−48.8	−48.6
$x_{13}(20)$ +50	+1.8	−13.60	−15.30	+21.30	−15.40	+27.40	+16.70	+21.8
−50	−1.8	+19	+18.40	−2.70	+18.50	−16.50	−13.80	−16.50

*Quantities within parentheses indicate value of design variables and behaviour quantities at initial points; further, $x_2 - x_1x$ represents, in inches, thickness distribution of cover skins; $x_4 - x_3x$ and $x_8 - x_7x$ represent, in inches, thickness distribution of webs of spars and ribs; $x_4 - x_3x$ and $x_{10} - x_9x$ represent, in square inches, area distribution of flanges of spars and ribs; x_{11}, x_{12}, and x_{13} represent aspect ratio, thickness to chord ratio, and sweep back angle, respectively

Table 16.3   Optimization Results for Minimum Weight Design Problem (using linearly approximated reanalysis whenever $\Delta x_i < 0.1$)

Behaviour quantities and design variables	Lower bound	Upper bound	Initial design	Sequential unconstrained minimization		
				r_1 (2.9)	r_2 (0.29)	r_3 (0.029)
δ_{tip} (in)		69	46	37.60	36.34	36.34
σ_{root} (ksi)		75	42	43.80	49.31	49.31
σ_{tip} (ksi)		75	8.03	8.66	9.82	9.82
$\sigma_{b\,root}$ (ksi)		55	41.9	43.70	49.19	49.19
$\sigma_{b\,tip}$ (ksi)		55	7.31	7.97	8.30	8.30
ω_1 (cps)	1.2	4.5	2.30	2.96	3.45	3.45
ω_2 (cps)	2	7.5	3.80	4.91	5.23	5.23
x_1	0.01	0.06	0.035	0.0259	0.0193	0.0193
x_2	0.10	0.60	0.35	0.2590	0.1925	0.1915
x_3	0.10	0.40	0.20	0.1988	0.1750	0.1750
x_4	0.10	0.50	0.35	0.1960	0.2790	0.2785
x_5	0.10	0.40	0.20	0.1988	0.1750	0.1750
x_6	0.10	0.50	0.35	0.1960	0.2790	0.2790
x_7	0.75	3	1.50	1.491	1.3120	1.3120
x_8	0.01	0.05	0.015	0.0189	0.0190	0.0190
x_9	10	35	20	21.40	21.50	21.50

Penalty function			1,03,500	87,000	34,500	29,500
Weight (lb)			51,900	38,300	29,100	28,990
Number of one-dimensional minimizations for each unconstrained optimization			9	4	2	
Total time			about 40 minutes			

Table 16.4 Optimization Results for Minimum Weight Design Problem (using exact reanalyses in third and fourth unconstrained minimizations)

Behaviour quantities and design variables	Lower bound	Upper bound	Initial design	Sequential unconstrained minimization			
				r_1 (2.9)	r_2 (0.29)	r_3 (0.029)	r_4 (0.0029)
δ_{tip} (in)		69	46	37.6	41.7	41.7	41.7
σ_{root} (ksi)		75	42	43.8	49.31	54.4	54.4
σ_{tip} (ksi)		75	8.03	8.66	9.82	10.5	10.5
$\sigma_{b\,root}$ (ksi)		55	41.9	43.7	49.19	54.2	54.2
$\sigma_{b\,tip}$ (ksi)		55	7.31	7.97	8.30	8.96	8.96
ω_1 (cps)	1.2	4.5	2.30	2.96	3.45	3.24	3.24
ω_2 (cps)	2	7.5	3.80	4.91	5.23	5.03	5.03
x_1	0.01	0:06	0.035	0.0259	0.0193	0.0188	0.0188
x_2	0.10	0.60	0.35	0.2590	0.1925	0.1880	0.1880
x_3	0.10	0.40	0.20	0.1988	0.1750	0.1818	0.1818
x_4	0.10	0.50	0.35	0.1960	0.2990	0.2780	0.2780
x_5	0.10	0.40	0.20	0.1988	0.1750	0.1820	0.1820
x_6	0.10	0.40	0.35	0.1960	0.2790	0.2780	0.2780
x_7	0.75	3	1.50	1.491	1.3120	1.368	1.368
x_8	0.01	0.05	0.015	0.0189	0.0190	0.0191	0.0191
x_9	10	35	20	21.40	21.50	20.20	20.20

Penalty function			1,03,500	87,000	34,500	29,100	28,380
Weight (lb)			51,900	38,300	29,100	28,300	28,300
Number of one-dimensional minimizations for each unconstrained optimization			9	4	8	2	
Total time			about 60 minutes				

REFERENCES

1   Schmit, L. A. and Thornton, W. A., Synthesis of an Airfoil at Supersonic Mach Number, NASA, CR-144, 1965.
2   Stroud, W. J., Dexter, C. B., and Stein, M., Automating Preliminary Design of Simplified Wing Structures to Satisfy Strength and Flutter Requirements, LWP-961, Langley Research Centre, Hampton, 1971.
3   Rao, S. S., Automated Optimum Design of Aircraft Wings to Satisfy Strength, Stability, Frequency and Flutter Requirements, RN 49, Div. Solid Mech., Struct. and Mech. Design, Case Western Reserve University, Ohio, 1971.
4   Giles, G. L., Procedure for automating aircraft wing structural design, *Proc. ASCE, J. Struct. Div.*, **97**, ST1, 99–113, 1971.
5   Rao, V. R., A study on the minimum weight design of wing structures satisfying strength, stability and frequency requirements, M. Tech. Thesis, Department of Aeronautical Engineering, IIT, Kanpur, 1976.

17
Optimization under Random Vibration Environment

17.1 INTRODUCTION

A large class of structural systems operate in random vibration environment. Such environment may be generated by (a) the acoustic pressure field of jet noise or boundary layer noise, (b) the ground motion during earthquakes, (c) atmospheric turbulence, (d) ground-induced vibrations in vehicles, and (e) ocean waves. The resulting random excitation makes the probabilistic approach essential in the design of structures.

The limitations of the conventional design procedures based on the factor of safety are well known. The reliability-based method, on the other hand, deals realistically with the uncertainty inherent in most design problems. (For a comprehensive review of the significant developments in structural reliability, see Ang[1].) Most of the work in this area has been confined to static problems.

Recently, attempts have been made by Shinozuka[2], Yang and Heer[3], Yang and Shinozuka[4], and Racicot[5] to extend the reliability-based design approach to dynamic problems. Nigam[6] and Narayanan[7] have applied the concepts of this design to structural optimization in random vibration environment. In this chapter, we shall give a general formulation of the structural optimization problem in random vibration environment within the reliability framework. We formulate this problem by considering the time dependence of the response quantities and then reducing it to a standard nonlinear programming (NLP) problem by eliminating the time dependence. In doing this, the weight of the structure is used as the optimality criterion, with constraints on the natural frequencies, buckling stresses, geometric dimensions, and dynamic responses such as stresses, acceleration, and fatigue life of the structure.

The constraints on the natural frequencies are important in designing dynamic systems. By keeping the natural frequencies away from frequencies where the excitation energy is large, the system responses can be reduced. It may be noted that human comfort and proper functioning of scientific payloads, among other considerations, may impose restrictions on the natural frequencies.

The constraints on dynamic responses in our formulation are expressed probabilistically. These constraints usually represent the failure or malfunctioning of the structure in a particular failure mode. Choosing the probability of failure in such a failure mode is a matter of engineering judgement based on the functions of the structural system and on the possible consequences. The probability of failure can be specified for the system as a whole or for each mode. We choose the latter for our study.

We shall apply the proposed formulation to (a) the optimum design of a cantilever beam with a tip mass; (b) an elevated water tank staging subjected to wind and earthquake loads;

and (c) sheet-stringer panels under jet noise excitation. The reduced NLP problems are solved by the sequential unconstrained minimization technique (SUMT), developed by Fiacco and McCormick[8], with an interior penalty function approach, together with the Fletcher-Powell method[9] (for the unconstrained optimization) which incorporates the golden section rule for the linear search.

17.2 FORMULATION

The structural optimization problem in random vibration environment is stated as: Find X to

$$\text{minimize } W(X) \tag{17.1}$$

subject to

$$P[\bigcup_{\substack{h=1 \\ 0 \leqslant t \leqslant T}}^{n} \{s_{hk}(Y(X, t)) \geqslant r_{hk}\}] \leqslant p_k \qquad (k = 1, 2, \ldots, m_1), \tag{17.2}$$

$$g_j(X) \leqslant \alpha_j \qquad (j = m_1 + 1, \ldots, m), \tag{17.3}$$

$$\omega_n^l \leqslant \omega_n(X) \leqslant \omega_n^u \qquad (n = 1, 2, \ldots, p), \tag{17.4}$$

where X is the vector of design variables, s_{hk} a function of the random dynamic response $Y(X, t)$, r_{hk} the deterministic or random limit on the function s_{hk}, p_k the specified probability of failure in the k-th failure mode, α_j the constraint on the deterministic function $g_j(X)$, ω_n^l and ω_n^u are respectively the specified lower and upper bounds on the n-th natural frequency of the system $\omega_n(X)$, $P[\ldots]$ denotes the probability of the event inside the square brackets, and $\bigcup$ the union of events.

In inequalities (17.2), the probability of failure is defined over a time interval $[0, T]$. This adds a dimension of difficulty to the optimization problem. To reduce the problem to a standard NLP problem, the probability of the union of events specifying failure must be evaluated. This is possible if acceptable upper bounds for the probabilities expressed by the left-hand side of the inequalities are established. Let

$$E_{hk} = [s_{hk}\{Y(X, t)\} \underset{0 \leqslant t \leqslant T}{\geqslant} r_{hk}], \tag{17.5}$$

$$P[E_{hk}] \leqslant q_{hk}(X, T). \tag{17.6}$$

Under the assumptions (17.5) and (17.6), inequalities (17.2) can be replaced by

$$P[\bigcup_{h=1}^{n} E_{hk}] \leqslant p_k \qquad (k = 1, 2, \ldots, m_1). \tag{17.7}$$

Bounds on either side of the probability

$$P[\bigcup_{h=1}^{n} E_{hk}]$$

can be established by considering the nature of the individual events E_{hk}. When each E_{hk} is independent and the q_{hk}'s are small, a close upper bound for

$$P[\bigcup_{h=1}^{n} E_{hk}]$$

can be obtained as

$$P[\bigcup_{h=1}^{n} E_{hk}] \leqslant \sum_{h=1}^{n} q_{hk}(X, T). \tag{17.8}$$

If the E_{hk}'s are fully correlated, then we have

$$[P \bigcup_{h=1}^{n} E_{hk}] \leqslant \max_{h} [q_{hk}(X, T)]. \tag{17.9}$$

It can also be shown that

$$P[\bigcup_{h=1}^{n} E_{hk}] \leqslant q_{1k} + a_h q_{hk}, \tag{17.10}$$

$$\bar{E}_{hk} = [s_{hk}\{\underset{0 \leqslant t \leqslant T}{Y(X, t)}\} < r_{hk}],$$

$$a_h = P[\bar{E}_{1k} \cap \bar{E}_{2k} \cap \ldots \cap \bar{E}_{h-1, k} \,|\, \bar{E}_{hk}], \qquad h = 2, 3, \ldots, n, \tag{17.11}$$

where a_h is defined in terms of the survival events and its values depend on the order of the component events in the particular failure mode. The evaluation of relation (17.11) would involve multiple integrations even if the joint distributions needed were available. A useful set of upper bounds for the conditional probability s_h can be obtained when all except one of the survival events are eliminated. From (17.11), we get

$$a_h \leqslant P[\bar{E}_{lk} \,|\, E_{hk}] = a_{hl} \qquad (l = 1, 2, \ldots, h - 1). \tag{17.12}$$

The closest upper bound to a_h is obtained by choosing the index l which minimizes the set of upper bounds. Choose

$$a_h^* = \min_{l=1}^{h-1} (a_{hl}). \tag{17.13}$$

Vanmarcke[10] has given an approximate result for the probabilities a_{hl} in terms of the correlation coefficients of the events. Thus,

$$a_{hl} \approx 1 - P[s_{hk}\{Y(X, t)\} \geqslant \max (r_{hk}, r_{hk} - \mu_{hk}^{M} + \frac{\sigma_{hk}^{M}\mu_{lk}^{M}}{\sigma_{lk}^{M}|\rho_{hl}^{M}|})]/P[s_{hk}\{Y(X, t)\} \geqslant r_{hk}]. \tag{17.14}$$

The quantities in relation (17.14) are defined in terms of the statistics of the safety margin $M_{ik} = r_{ik} - s_{lk}$. μ_{ik}^{M} is the mean of M_{ik}, σ_{ik}^{M} the standard deviation, and ρ_{ij}^{M} the correlation coefficient between the random variables M_i and M_j.

From relations (17.13) and (17.14), a close upper bound a_h^* can be determined. On substituting (17.14) in Eq. (17.10), we get

$$P[\bigcup_{h=1}^{n} E_{hk}] \leqslant \sum_{h=1}^{n} a_h^* q_{hk}, \tag{17.15}$$

where

$$a_1^* = 1,$$

$$a_2^* = a_2 = P[\bar{E}_{1k} \,|\, E_{2k}],$$

$$a_3^* = \min (P[\bar{E}_{1k} \,|\, E_{3k}], P[E_{2k} \,|\, E_{3k}]),$$

and so on.

Thus, (17.2) can be replaced by any one of the inequalities

$$\sum_{h=1}^{n} [q_{hk}(X, T)] \leqslant p_k \qquad (k = 1, 2, \ldots, m_1), \tag{17.16}$$

$$\max_{h=1}^{n} [q_{hk}(X, T)] \leqslant p_k \qquad (k = 1, 2, \ldots, m_1), \tag{17.17}$$

$$\sum_{h=1}^{n} [a_h^*(X)q_{hk}(X, T)] \leqslant p_k \qquad (k = 1, 2, \ldots, m_1). \tag{17.18}$$

The structural optimization problem in random vibration environment can now be stated in the standard NLP format: Find X^* to

$$\text{minimize } W(X) \tag{17.1}$$

subject to

$$\sum_{h=1}^{n} [q_{hk}(X, T)] \leqslant p_k \qquad (k = 1, 2, \ldots, m_1) \tag{17.16}$$

or

$$\max_{h=1}^{n} [q_{hk}(X, T)] \leqslant p_k \qquad (k = 1, 2, \ldots, m_1) \tag{17.17}$$

or

$$\sum_{h=1}^{n} [a_h^*(X)q_{hk}(X, T)] \leqslant p_k \qquad (k = 1, 2, \ldots, m_1) \tag{17.18}$$

and

$$g_j(X) \leqslant \alpha_j \qquad (j = m_1 + 1, \ldots, m), \tag{17.3}$$

$$\omega_n^l \leqslant \omega_n(X) \leqslant \omega_n^u \qquad (n = 1, 2, \ldots, p). \tag{17.4}$$

In structural systems, failure can be specified in terms of (a) the first excursion above a given level, (b) the fraction of time spent above the specified level, and (c) the cumulative damage level, as in fatigue. In the formulation just given, the first two modes of failure are specified by (17.2) and reduced to (17.16) or (17.17) or (17.18). Fatigue failure is expressed in terms of the expected rate of fatigue damage which is determined on the basis of the Palmgren-Miner cumulative damage hypothesis and specified by inequalities (17.3). In the class of problems where the response can be treated as stationary and the time interval $(0, T)$ is large, it is possible to simplify probability estimates by assuming time independence of response statistics.

The failure of the structure may arise also from buckling. If so, the behaviour constraints may include the buckling constraints. Sometimes it is necessary to satisfy the assumptions made in the structural analysis by specifying the constraints on the relative dimensions of the structural system. These and other side constraints on the design variables are included in (17.3).

Frequency is an important parameter in relation to human comfort, proper performance of scientific payloads, and equipment in flight vehicles. Large responses associated with resonant conditions can be avoided by imposing suitable constraints on natural frequencies which are of special significance in designing systems for dynamic loads. In random vibration problems, if the power spectral density (psd) is flat over the frequency range of interest, response calculations can be simplified by locally assuming white noise for the excitation and the frequency constraints expressed by (17.4).

17.3 OPTIMIZATION TECHNIQUE

The foregoing NLP problem is solved by using the sequential unconstrained minimization technique of Fiacco and McCormick[8], together with the Fletcher-Powell algorithm[9] for the unconstrained optimization. The linear search for the step length along any direction is conducted by adopting the golden section rule. The interior penalty function used for unconstrained minimization is given by

$$P(X, r) = W(X) + r \sum_i \frac{1}{g_i(X)}. \tag{17.19}$$

The initial value of r is so chosen as to weigh the objective function and the penalty terms equally at the starting point. That is,

$$r_1 = W(X_0)/\sum_i \frac{1}{g_i(X_0)}, \tag{17.20}$$

where X_0 is the initial design vector. When this vector is near one or more of the constraint surfaces, the value of r_1 as given by Eq. (17.20) tends to be small, thus creating computation difficulties. In such cases, r_1 is chosen as

$$r_1 = W(X_0)/[\sum_i \frac{1}{g_i(X_0)} + \sum_k \{5 - \frac{1}{h_k(X_0)}\}], \tag{17.21}$$

where k is the set of constraints for which $g_k(X_0) \leqslant 0.2$. The summation indicated by the index i is carried through all the constraints.

The sequential unconstrained minimization algorithm is terminated when the convergence criteria are satisfied in the following order:

$$\frac{P^*(X, r) - W^*(X)}{W^*(X)} \leqslant \epsilon_1, \tag{17.22}$$

$$\frac{P^*(X, r_i) - P^*(X, r_i)_1}{P^*(X, r_i)_1} \leqslant \epsilon_2,$$

$$|X_{i+1}^* - X_i^*| \leqslant \epsilon_3, \tag{17.23}$$

$$i > p, \tag{17.24}$$

where $P^*(X, r)$ is the minimum value of the P-function at the end of an unconstrained minimization cycle, $W^*(X)$ the corresponding value of the objective function, and the subscript i refers to the number of the sequence of unconstrained minimizations. The constants chosen in the convergence criterion are $\epsilon_1 = 10^{-4}$, $\epsilon_2 = 5 \times 10^{-4}$, $\epsilon_3 = 10^{-4}$, and $p = 10$.

An extrapolation technique is adopted at the end of two P-function minimizations to locate better starting points for subsequent unconstrained minimizations. Let X_i^*, X_{i+1}^* be the design vectors that minimize the functions $P(X, r_i)$, $P(X, r_{i+1})$, respectively. Then a good starting point for the $(i + 2)$-th P-function minimization is

$$X_{i+2} = X_{i+1} + \delta^{1/2}(X_{i+1}^* - X_i^*), \tag{17.25}$$

where δ is the constant which reduces the parameter r at the end of each P-function minimization. That is,

$$r_{i+1} = \delta r_i \qquad (\delta < 1.0). \tag{17.26}$$

In the problems we shall now consider, the value of δ is taken as 0.1. Figure 17.1 shows the flowchart for optimization.

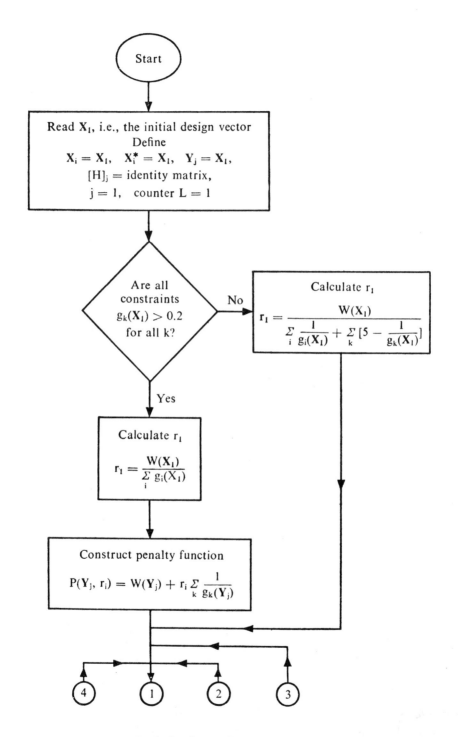

Fig. 17.1   Flowchart for Optimization under Random Vibration Environment (cont.).

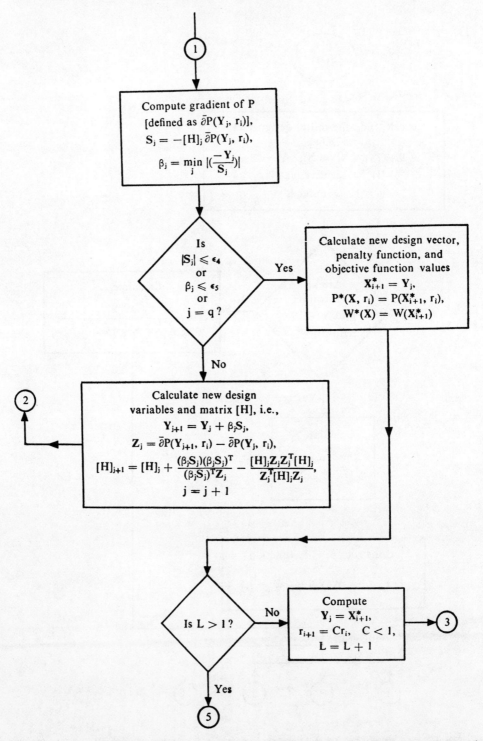

Fig. 17.1   Flowchart for Optimization under Random Vibration Environment (cont.).

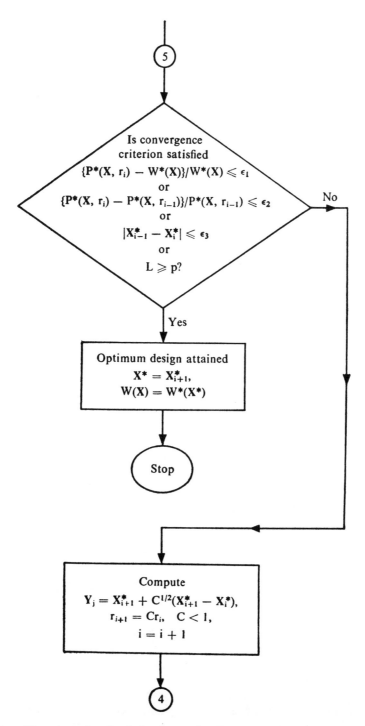

Fig. 17.1   Flowchart for Optimization under Random Vibration Environment.

17.4 EXAMPLES

17.4.1 CANTILEVER BEAM WITH TIP MASS

Consider a beam of thin-walled rectangular box section supporting a tip mass (see Fig. 17.2). The optimization problem is formulated for determining the optimum dimensions of the cross-section, i.e., the width, depth, and thickness of the beam, such that the beam is of minimum

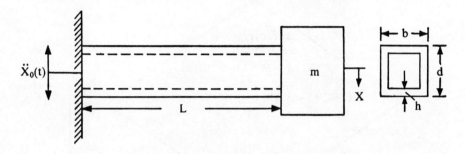

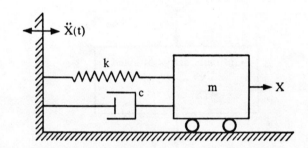

17.2   Cantilever Beam with Tip Mass and Its Idealized Model.

weight satisfying the constraints to be specified. The mass of the beam is assumed to be negligible compared with that at the tip. On this assumption, the beam can be idealized as a single-degree-of-freedom system. It is subjected to a Gaussian white noise excitation of constant psd Φ_0 at the support.

Objective Function

The design is defined in its totality by the design variable vector X, given by

$$X^T = \{b, d, h\}^T, \tag{17.27}$$

that is, $x_1 = b$, $x_2 = d$, and $x_3 = h$. Since the beam is of uniform cross-section and fixed length L, minimizing the weight is equivalent to minimizing the area of cross-section of the beam. Thus, the objective function becomes

$$W(X) = 2x_3(x_1 + x_2), \tag{17.28}$$

assuming that the thickness of the beam h is much smaller than the breadth b and depth d of the cross-section.

Frequency Constraints

The natural frequency of a cantilever beam is required to be greater than a lower bound. So the frequency constraint is expressed in the form

$$\omega(X) \geqslant \omega^l \tag{17.29}$$

(where ω^l is the specified lower limit on the natural frequency) and, in a normalized form, it becomes

$$g_1(X) = 1 - \frac{\omega^l}{\omega(X)} \geqslant 0. \tag{17.30}$$

For the single-degree-of-freedom system, we obtain

$$\omega(X) = \left(\frac{3EI}{mL^3}\right)^{1/2}, \tag{17.31}$$

where I is the second moment of area of the cross-section, m the tip mass, and E Young's modulus. In terms of the design variables, the moment of inertia is given by

$$I = \tfrac{1}{12}(x_1 x_2^2 x_3 - 12 x_1 x_2 x_3^2 + 2 x_2^3 x_3 - 12 x_2^2 x_3^2). \tag{17.32}$$

Fatigue Damage Constraint

The constraint on the fatigue life of the structure is expressed in terms of the expected rate of fatigue damage which is based on the linear cumulative damage law of Miner[11]. This rate should be less than a specified value, that is,

$$\frac{1}{T}E[D(X)] \leqslant D_1 \tag{17.33}$$

or, in a normalized form,

$$g_2(X) = 1 - \frac{v_0^*}{\beta D_1}(\sqrt{2}\sigma_s)^\alpha \Gamma(1 + \frac{\alpha}{2}) \geqslant 0, \tag{17.34}$$

where v_0^* is the rate of zero crossings, σ_s the standard deviation of the maximum stress at the root of the cantilever, $\Gamma(\ldots)$ the gamma function, α and β are the constants in the fatigue damage law given by $Ns^\alpha = \beta$, $E[D(X)]$ is the expected rate of fatigue damage, and

$$\sigma_s = \frac{3Ex_2}{2L}\left(\frac{\pi\rho_0}{2\xi\omega^3}\right)^{1/2}, \tag{17.35}$$

ξ being the damping factor.

Stress Constraint

The stress constraint in our problem is expressed in terms of the first excursion failure probability which is evaluated in two ways. First, the crossings of the stress response over the specified level are assumed to be independent; on this assumption, the first excursion failure probability is given by

$$p_f(X) = 1.0 - \exp\left[-\frac{\omega T}{\pi}\exp\{-\sigma_y^2/(2\sigma_s^2)\}\right], \tag{17.36}$$

where σ_y is the yield stress which is assumed to be deterministic and T the time of operation of the random process. Second, the two-state zero-one process formed by the stress response crossings is assumed to be the Markov process. On this assumption which is due to Vanmarcke[12], the first excursion failure probability is obtained as

$$p_f(X) = 1.0 - \exp\left[-\frac{\omega T}{\pi}\frac{1 - \exp(-\sigma_y q\sqrt{\pi/2}/\sigma_s)}{\exp(\frac{1}{2}\sigma_y^2/\sigma_s^2) - 1}\right], \tag{17.37}$$

where q is a shape factor of the spectral density of the stress response given, in terms of the spectral moments λ_i, as

$$q = (1 - \frac{\lambda_1^2}{\lambda_0\lambda_2})^{1/2}. \tag{17.38}$$

The λ_i's are defined by

$$\lambda_i = \int_0^\infty \omega^i G_s(\omega)\, d\omega, \tag{17.39}$$

where $G_s(\omega)$ is the one-sided psd of the stress response.

The stress constraint can now be stated in the form

$$P[\,\substack{|s(t)| \\ 0 \leqslant t \leqslant T} \geqslant \sigma_y] \leqslant p_1. \tag{17.40}$$

In the normalized form, it is expressed as

$$g_3(X) = 1 - p_f(X)/p_1 \geqslant 0, \tag{17.41}$$

where $p_f(X)$ is given by either Eq. (17.36) or Eq. (17.37).

Acceleration Constraint

The acceleration constraint is expressed in the form

$$P[\,\substack{|a(t)| \\ 0 \leqslant t \leqslant T} \geqslant a_1] \leqslant p_2, \tag{17.42}$$

where $a(t)$ is the absolute acceleration of the mass.

Assuming the random process $a(t)$ to be stationary ($T \to \infty$), we write inequality (17.42) as

$$g_4(X) = 1 - \text{erfc}\,[a_1/(\sqrt{2}\sigma_a)]/p_2 \geqslant 0, \tag{17.43}$$

where erfc (...) is the complementary error function and σ_a the standard deviation of the absolute acceleration of the mass. Inequality (17.43) also means that the expected fraction of time that the acceleration response $|a(t)|$ spends above the critical level a_1 may be less than the specified fraction p_2.

Buckling and Side Constraints

The local buckling constraints on the web and flange of the box beam are expressed in terms of the buckling coefficients and are defined as

$$g_5(X) = 1 - \sigma_y/[K_1E(x_3/x_2)^2] \geqslant 0, \tag{17.44}$$

$$g_6(X) = 1 - \sigma_y/[K_2E(x_3/x_1)^2] \geqslant 0. \tag{17.45}$$

In order that the Euler beam assumption is valid in our analysis, the side constraints are expressed as

$$g_7(X) = 1 - x_1/(L/10) \geqslant 0, \tag{17.46}$$

$$g_8(X) = 1 - x_2/(L/10) \geqslant 0. \tag{17.47}$$

In addition, the non-negativity restrictions on the design variable vector lead to more constraints, namely,

$$x_1 \geqslant 0, \tag{17.48}$$

$$x_2 \geqslant 0, \tag{17.49}$$

$$x_3 \geqslant 0. \tag{17.50}$$

Optimization Problem Statement

The structural optimization problem for a cantilever beam with tip mass can now be stated as follows. Find X^* to

$$\text{minimize } W(X) = 2x_3(x_1 + x_2) \tag{17.51}$$

subject to

$$g_1(X) = 1 - \frac{\omega^l}{\omega(X)} \geqslant 0,$$

$$g_2(X) = 1 - \frac{v_0^*}{\beta D_1}(\sqrt{2}\sigma_s)^\alpha \Gamma(1 + \frac{\alpha}{2}) \geqslant 0,$$

$$g_3(X) = 1 - p_f(X)/p_1 \geqslant 0,$$

$$g_4(X) = 1 - \text{erfc } [a_1/(\sqrt{2}\sigma_a)]/p_2 \geqslant 0,$$

$$g_5(X) = 1 - \sigma_y/[K_1 E(x_3/x_2)^2] \geqslant 0,$$

$$g_6(X) = 1 - \sigma_y/[K_2 E(x_3/x_1)^2] \geqslant 0, \tag{17.52}$$

$$g_7(X) = 1 - x_1/(L/10) \geqslant 0,$$

$$g_8(X) = 1 - x_2/(L/10) \geqslant 0,$$

$$g_9(X) = x_1 \geqslant 0,$$

$$g_{10}(X) = x_2 \geqslant 0,$$

$$g_{11}(X) = x_3 \geqslant 0.$$

We use the following data to obtain the results for the optimization problem just stated:

$$\Phi_0 = 0.02 \ (\text{ft/sec}^2)^2/(\text{cycle/sec}),$$

$$m = 1.0 \ \text{lb sec}^2/\text{in},$$

$$\xi = 0.02,$$

$$L = 200 \ \text{in},$$

$$\omega^l = 5 \ \text{cps},$$

$$p_2 = 10^{-3},$$

$$a = 3864 \ \text{in/sec}^2,$$

$$\sigma_y = 20{,}000 \ \text{lb/in}^2,$$

$$E = 10.6 \times 10^6 \ \text{psi},$$

$\alpha = 6.0,$

$\beta = 6.4 \times 10^{31},$

$K_1 = 21.72,$

$K_2 = 3.62,$

$T = 3600$ sec,

two values of p_1 at the constant fatigue damage level D_1,

two values of D_1 at the constant value of p_1.

The results are given in Tables 17.1, 17.2, and 17.3 for the assumption of both the Poisson crossing rate [see Eq. (17.36)] and the Vanmarcke failure rate [see Eq. (17.37)]. These results are plotted in Figs. 17.3 and 17.4 which show the convergence to the optimum from two different initial points. From the results obtained, the following observations can be made:

(i) The difference in the optimal weights of the cantilever beam based on Poisson's and Vanmarcke's assumptions regarding the first excursion failure is not significant; nevertheless, Poisson's assumption is conservative.

(ii) A change in the probability of failure p_1 from 10^{-3} to 10^{-4} causes 9.2 per cent increase in the optimal weight.

(iii) The fatigue damage constraint becomes active for the damage rate $D_1 = 5 \times 10^{-9}$, raising the optimal weight by 2 per cent.

Table 17.1   Cantilever Beam with Tip Mass
Given

$$p_1 = 10^{-3}, \qquad D_1 = 5 \times 10^{-8}.$$

	r	x_1	x_2	x_3	$W(X)$	$W(X) + r \sum_i \frac{1}{g_i(X)}$
Initial design	0.8451	10	18	0.3	16.80	37.825
Optimum design (Poisson crossing rate)	0.8451×10^{-9}	8.43	19.992	0.192	10.944	10.944
Optimum design (Vanmarcke failure rate)	0.8451×10^{-8}	8.349	19.997	0.191	10.808	10.810
Initial design	0.9019	14	14	0.35	19.6	40.122
Optimum design (Poisson crossing rate)	0.9019×10^{-2}	8.465	19.874	0.193	10.959	10.960
Optimum design (Vanmarcke failure rate)	0.9019×10^{-7}	8.351	19.991	0.191	10.809	10.810

Active constraints at optimum g_3, g_6, g_8

Table 17.2   Cantilever Beam with Tip Mass
Given

$$p_1 = 10^{-4}, \qquad D_1 = 5 \times 10^{-8}.$$

	r	x_1	x_2	x_3	$W(X)$	$W(X) + r \sum_i \dfrac{1}{g_i(X)}$
Initial design	0.8451	10	18	0.3	16.8	37.826
Optimum design (Poisson crossing rate)	0.8451×10^{-8}	8.995	19.677	0.205	11.781	11.784
Optimum design (Vanmarcke failure rate)	0.8451×10^{-9}	9.093	19.174	0.207	11.737	11.738
Initial design	1.1881	14	14	0.4	22.4	44.8
Optimum design (Poisson crossing rate)	1.1881×10^{-5}	9.026	19.580	0.206	11.796	11.797
Optimum design (Vanmarcke failure rate)	1.8881×10^{-6}	9.182	18.893	0.209	11.782	11.785

Active constraints at optimum g_3, g_6

Table 17.3   Cantilever Beam with Tip Mass
Given

$$p_1 = 10^{-4}, \qquad D_1 = 5 \times 10^{-9}.$$

	r	x_1	x_2	x_3	$W(X)$	$W(X) + r \sum_i \dfrac{1}{g_i(X)}$
Initial design	0.89502	14	14	0.35	19.6	40.11
Optimum design (Poisson crossing rate)	0.8952×10^{-9}	9.143	19.816	0.209	12.093	12.101
Optimum design (Vanmarcke failure rate)	0.8952×10^{-8}	9.126	19.876	0.208	12.089	12.090
Initial design	0.8293	10	18	0.3	16.8	37.746
Optimum design (Poisson crossing rate)	0.8293×10^{-6}	9.099	19.897	0.209	12.100	12.193
Optimum design (Vanmarcke failure rate)	0.8293×10^{-8}	9.094	19.992	0.207	12.081	12.082

Active constraints at optimum g_2, g_5, g_8

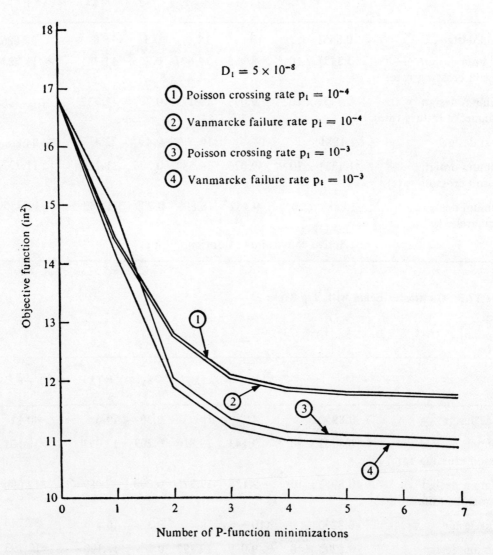

Fig. 17.3 Sequence of Unconstrained Minima (Cantilever beam with tip mass).

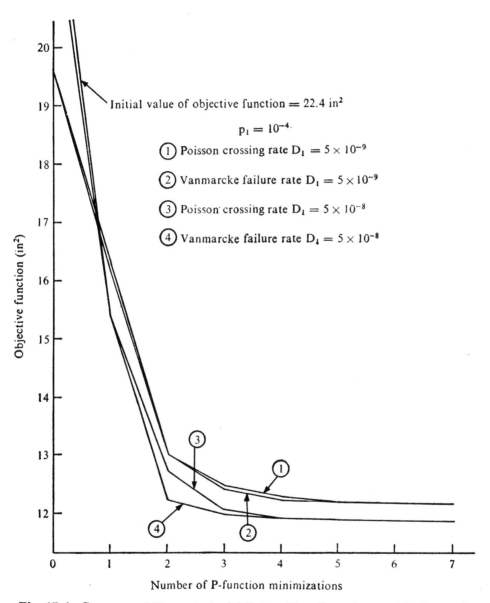

Fig. 17.4 Sequence of Unconstrained Minima (Cantilever beam with tip mass).

17.4.2   ELEVATED WATER TANK STAGING (Earthquake loads)

Consider a truss structure supporting a water tank, as shown in Fig. 17.5a. The base of the structure is subjected to ground acceleration during an earthquake. For the response analysis, the structure is idealized as a single-degree-of-freedom system (see Fig. 17.5b). The stiffness of

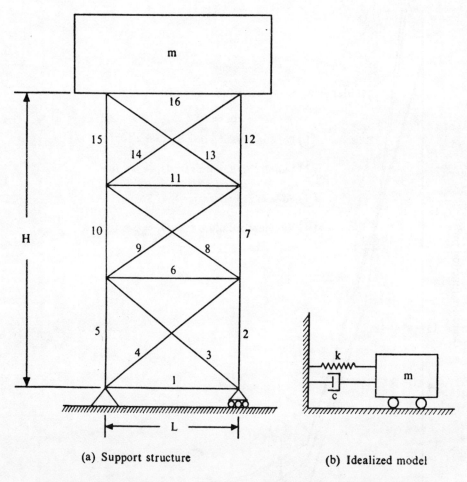

(a) Support structure   (b) Idealized model

Fig. 17.5   Support Structure for Water Tank and Its Idealized Model.

the system is computed by using the flexibility analysis. Following Pestel and Leckie[13], we obtain the deflections δ and the member force $\{f\}$ of the truss as

$$\delta = [D_{00} - D_{10}^T D_{11}^{-1} D_{10}]\{p\}, \tag{17.53}$$

$$\{f\} = [B_0 - B_1 D_{11}^{-1} D_{10}]\{p\}, \tag{17.54}$$

where $[B_0]$ is the matrix relating the member forces of a reduced statically determinate structure to the external force vector $\{p\}$, $[B_1]$ the matrix operating on the redundant force vector which gives additional member forces due to redundants, and $[D_{10}]$, $[D_{11}]$ are the matrices obtained by operations on $[B_0]$, $[B_1]$ and on the element flexibility matrix of the structure.

Objective Function

The objective function is the total volume of the structure, i.e.,

$$W(X) = \sum_{i=1}^{16} A_i L_i, \tag{17.55}$$

where A_i is the area of cross-section of the i-th member and L_i the length of the i-th member. There are 12 design variables in the problem given by the vector

$$X = \{x_1, x_2, \ldots, x_{12}\}^T$$
$$= \{A_1, A_2, A_3, A_6, A_7, A_8, A_{11}, A_{12}, A_{13}, L_2, L_7, L_{12}\}^T. \tag{17.56}$$

From symmetry, the relations assumed valid are:

$$A_2 = A_5, \qquad A_3 = A_4, \qquad A_7 = A_{10}, \qquad A_8 = A_9, \qquad A_{12} = A_{15}, \qquad A_{13} = A_{14};$$

$$L_2 = L_5, \qquad L_3 = L_4, \qquad L_7 = L_{10}, \qquad L_8 = L_9, \qquad L_{12} = L_{15}, \qquad L_{13} = L_{14}.$$

Also

$$L_1 = L_6 = L_{11} = L_{16} = L, \qquad L_{12} = H - (L_2 + L_7),$$

where L is the base width and H the total height of the truss. The lengths of the vertical members of the truss are treated as design variables with the condition that the sum of their lengths equals H. Instead of keeping this condition as a constraint in the problem, namely,

$$L_2 + L_7 + L_{12} = H,$$

only the two lengths L_2 and L_7 are considered design variables in Eq. (17.56) so that $L_{12} = H - (L_2 + L_7)$ is automatically determined.

Frequency Constraint

The constraint on the natural frequency of the structure is so specified that when the tank is partially filled the first few frequencies of liquid oscillations are kept well below the natural frequency of the structure. This is done to avoid large amplitude liquid sloshing due to earthquake excitation. The frequencies of liquid oscillations are calculated by using the formulae given by Housner[14]. The natural frequency is $\omega_0(X) = \sqrt{k/m}$, where k is the stiffness of the truss and m the total mass of the water tank. The frequency constraint is expressed as

$$\omega^l \leqslant \omega_0(X). \tag{17.57}$$

Stress Constraint

The ground acceleration during an earthquake is assumed to be a stationary random process with one-sided psd (see Housner and Jennings[15]) expressed in terms of $(ft/sec^2)^2/(rad/sec)$ as

$$G(\omega) = C(1 + \omega^2/147.8)/[(1 - \omega^2/242)^2 + \omega^2/147.8], \tag{17.58}$$

where C is a constant taken equal to 1.04 to correspond to the earthquake El-Centro 1940. For response calculation, the ground acceleration is locally considered white noise, the psd being given by Eq. (17.58).

The mean square value of the absolute acceleration of the water tank is defined by

$$\sigma_{\ddot{x}}^2 = \pi G(\omega_0)(1 + 4\zeta^2)/(4\zeta), \tag{17.59}$$

where ζ is the critical damping ratio. The standard deviation of the lateral force exerted by the water tank on the support structure is given by

$$\sigma_p = m\sigma_{\ddot{x}} \tag{17.60}$$

and the standard deviation of the stresses in the truss members by

$$\sigma_{si} = [B]\sigma_p/A_i, \tag{17.61}$$

where

$$[B] = [B_0 - B_1 D_{11}^{-1} D_{10}].$$

The dependence of the statistics in Eqs. (17.58)–(17.61) on the design variables is implicit. The excitation is assumed to be Gaussian. The response also is Gaussian since the system is linear. The stress constraint can be expressed in probabilistic form as

$$P[\overline{\bigcup_{\substack{i=1 \\ 0 \leqslant t \leqslant T}}^{16} \{s_i(X, t) - \sigma_y\}] \leqslant p_1, \tag{17.62}$$

where s_i is the stress in the i-th member and the yield stress σ_y is assumed to be deterministic. The probability of failure as expressed in inequality (17.62) implies the weakest link hypothesis.

The probability on the left-hand side of (17.62) is calculated on the basis of (a) the fraction of time the response spends above the yield stress level and (b) the first excursion failure probability. Assuming (17.62) is replaced by (17.16) or (17.17), we have, for (a),

$$\sum_{i=1}^{16} \text{erfc} \left[\sigma_y/(\sqrt{2}\sigma_{si})\right] \leqslant p_1 \tag{17.63}$$

or

$$\max_i \left[\text{erfc} \{\sigma_y/(\sqrt{2}\sigma_{si})\}\right] \leqslant p_1 \tag{17.64}$$

and, for (b),

$$\sum_{i=1}^{16} \left[1 - \exp\{-\frac{\omega T}{\pi} \exp\left(-\sigma_y^2/(2\sigma_{si}^2)\right)\}\right] \leqslant p_1 \tag{17.65}$$

or

$$\max_i \left[1 - \exp\{-\frac{\omega T}{\pi} \exp\left(-\sigma_y^2/(2\sigma_{si}^2)\right)\}\right] \leqslant p_1. \tag{17.66}$$

Acceleration Constraint

The acceleration constraint is expressed, in terms of the fractional occupation time of the absolute acceleration response of the mass exceeding a given value, as

$$\text{erfc} \left[a/(\sqrt{2}\sigma_{\ddot{x}})\right] \leqslant p_2, \tag{17.67}$$

where a is a specified upper bound on the absolute acceleration level of the mass.

Non-Negativity Restrictions

Apart from the constraints just considered, the non-negativity restrictions of the design variables require that

$$x_i \geqslant 0 \qquad (i = 1, 2, \ldots, 12), \tag{17.68}$$

where x_i is the i-th element of the design variable vector.

Optimization Problem Statement

The optimization problem in aseismic design can be stated in the final form as: Find X to

$$\text{minimize } W(X) = \sum_{i=1}^{16} A_i L_i$$

subject to

$$g_1(X) = 1 - \omega^l/\omega_0(X) \geqslant 0,$$

$$g_2(X) = 1 - p_f(X)/p_1 \geqslant 0,$$

$$g_3(X) = 1 - \text{erfc}\,[a/(\sqrt{2}\sigma_{\bar{x}})]/p_2 \geqslant 0,$$

$$x_i \geqslant 0 \qquad (i = 1, 2, \ldots, 12),$$

where the constraints g_1–g_3 have been normalized to obtain values between 0 and 1. In the constraint g_2, $p_f(X)$ is the probability of failure in the stress constraint given by the left-hand side of any one of the inequalities (17.63), (17.64), (17.65), and (17.66).

The numerical results of optimization of the truss structure subjected to an earthquake ground acceleration are obtained for the following set of design data:

$m = 5160$ lb sec^2/in,

$H = 1080$ in,

$L = 360$ in,

$E = 30 \times 10^6$ lb/in^2,

$\sigma_y = 3000$ lb/in^2,

$\omega^l = 5$ rad/sec,

$p_1 = 10^{-3}$,

$p_2 = 10^{-3}$,

$\zeta = 0.02$,

$a = 360$ in/sec^2,

$T = 25$ sec.

Assuming that the tank has a square cross-section $20' \times 20'$ and height $20'$, we get the liquid sloshing frequencies

$\Omega_1 = 1.28$ rad/sec

$\Omega_2 = 2.15$ rad/sec   ($\frac{1}{2}$ full)

$\Omega_3 = 2.74$ rad/sec

$\Omega_1 = 1.45$ rad/sec

$\Omega_2 = 2.24$ rad/sec   ($\frac{3}{4}$ full)

$\Omega_3 = 2.76$ rad/sec

For details of the technique applied to obtain the optimum value, refer to Narayanan[7]. The results are given in Table 17.4 and the sequence of unconstrained optima converging to the optimal solution is shown in Fig. 17.6.

Table 17.4 Truss Structure (aseismic design)

	r	x_1	x_2	x_3	x_4	x_5	x_6	x_7	x_8	x_9	x_{10}	x_{11}	x_{12}	$W(X)$
Initial*	11.53	100	100	100	100	100	100	100	100	100	100	360	360	6,65,470
Optimal*	1.15×10^{-4}	38.3	46.5	56	142	32.1	45.5	52.8	21.4	28.4	20.6	383	357	2,97,346
Initial**	18.04	140	140	180	180	140	140	150	170	120	120	360	360	9,84,422
Optimal**	1.8×10^{-4}	56.6	58.9	70.4	136	49.9	54.7	77.7	33	36	49.5	334	362	3,78,740
Initial†	19.53	180	180	180	180	140	160	160	160	140	140	360	360	10,71,952
Optimal†	1.95×10^{-4}	93.4	65.9	92	122	60.3	78.1	88.7	36.7	50.1	66.5	313	382	4,71,571
Initial‡	12.88	180	120	120	150	120	100	150	120	100	140	360	360	8,08,234
Optimal‡	1.29×10^{-4}	144	86	108	135	75.5	83.6	117	49.8	65.3	107	131	335	5,89,539

r is penalty parameter in SUMT method
* Member stresses correlated with stress constraint in terms of mean fraction of time (see curve 1 in Fig. 17.6)
** Member stresses uncorrelated with stress constraint in terms of mean fraction of time (see curve 2 in Fig. 17.6)
† Member stresses correlated with stress constraint in terms of first excursion failure (see curve 3 in Fig. 17.6)
‡ Member stresses uncorrelated with stress constraint in terms of first excursion failure (see curve 4 in Fig. 17.6)

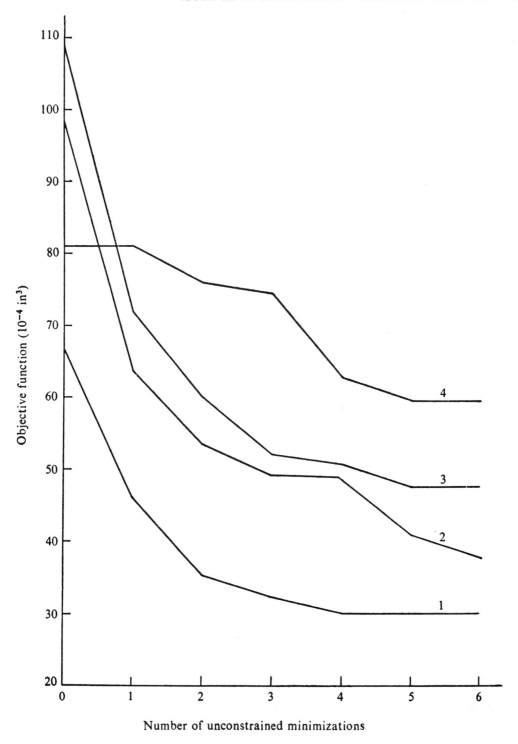

Fig. 17.6 Sequence of Unconstrained Optima (Truss structure, earthquake design).

We may draw the following observations from our study:

(i) Since the stress constraint is the only active constraint at the optimum point, the optimal design is sensitive to the way in which this constraint is specified. The optimal volume is higher when the constraint is expressed in terms of the first excursion failure. For instance, when we compare the results marked * and † in Table 17.4, we see that the optimal volume increases by 58 per cent if failure is expressed in terms of the first excursion failure. This is so because the first excursion failure probability based on independent crossing is an upper bound on the first excursion probability, whereas the failure expressed in terms of the mean fractional occupation time is a lower bound on it (see Shinozuka[16]).

(ii) The optimal design is sensitive to the degree of correlation between member stresses. Comparing the results marked † and ‡ in Table 17.4, we get an optimal volume less by 25 per cent for fully correlated member stresses.

17.4.3   ELEVATED WATER TANK STAGING (Wind loads)

The design of structures for an elevated water tank staging fits into our general formulation of the structural optimization problem in random vibration environment given in Section 17.2. (For a discussion on the random vibration analysis of wind-loaded structures, see Davenport[17] and Racicot[5].)

Let us consider the design of the truss structure (see Fig. 17.5a) against wind-induced loads. The governing differential equation for the response of the idealized single-degree-of-freedom structure is given by

$$m\ddot{x} + c\dot{x} + kx = f(t), \tag{17.69}$$

where x is the displacement at the top of the structure and $f(t)$ the force due to wind.

According to Davenport[17], neglecting the effect of unsteady oscillations, we get the force $f(t)$ as

$$f(t) = \tfrac{1}{2}A\rho C_D V_t |V_t|, \tag{17.70}$$

where ρ is the density of air, V_t the relative wind velocity dependent on height, C_D the drag coefficient, and A the frontal area.

Let $U(t)$ be the absolute velocity of the wind, that is,

$$U(t) = V_t + \dot{x}(t), \tag{17.71}$$

where $U(t)$ is assumed to be composed of a steady component V, called the *mean wind velocity* which is independent of time, and a fluctuating component $v(t)$, known as the *gust velocity*. Thus, Eq. (17.71) may be written as

$$U(t) = V + v(t). \tag{17.72}$$

From Eqs. (17.70) and (17.71), the force $f(t)$ due to wind action can, after neglecting second-order terms in $v(t)$ and $x(t)$ which are small, be obtained as

$$f(t) = \tfrac{1}{2}\rho A C_D[V^2 + 2Vv(t)\alpha - 2V\dot{x}]. \tag{17.73}$$

Substituting Eq. (17.73) in Eq. (17.69), we get

$$m\ddot{x} + (c + V\rho A C_D)\dot{x} + kx = \tfrac{1}{2}\rho A C_D V^2 + \rho A C_D V v(t). \tag{17.74}$$

If we assume that the response $x(t)$ of the structure due to wind is composed of a mean displacement Y and a random component $y(t)$, defined by

$$x = Y + y(t), \qquad (17.75)$$

the governing differential equation (17.74) becomes

$$m\ddot{y} + (c + \rho A C_D V)\dot{y} + ky = \rho A C_D V v(t), \qquad (17.76)$$

$$Y = \tfrac{1}{2}\rho A C_D V^2 / k. \qquad (17.77)$$

Equation (17.76) may be rewritten as

$$\ddot{y} + 2\zeta\omega_0\dot{y} + \omega_0^2 y = \frac{\rho A C_D V v(t)}{m}, \qquad (17.78)$$

where

$$\omega_0^2 = k/m, \qquad (\rho A C_D V + c)/m = 2\zeta\omega_0.$$

Thus, by knowing the statistics of the random component $v(t)$, we can obtain the statistics of $y(t)$, $\dot{y}(t)$, and $\ddot{y}(t)$ from random vibration analysis.

The two-sided psd of $v(t)$ is, according to Yang and Shinozuka[4], given by

$$\Phi_v(\omega) = \frac{2KV_1}{\pi}\left[\frac{X|\omega|/(\pi V_1)}{[1 + \{X\omega/(\pi V_1)^2\}]^{4/3}}\right], \qquad (17.79)$$

where K is the surface drag coefficient, V_1 the mean wind velocity at a reference height $Z = 33$ ft, $X = 2000$ ft, and ω the frequency in rad/sec. From random vibration theory, the psd's of y, $\dot{y}$, $\ddot{y}$ are respectively

$$\Phi_y(\omega) = |H(\omega)|^2\left(\frac{\rho A C_D V}{m}\right)^2\Phi_v(\omega),$$

$$\Phi_{\dot{y}}(\omega) = \omega^2\Phi_y(\omega), \qquad (17.80)$$

$$\Phi_{\ddot{y}}(\omega) = \omega^4\Phi_y(\omega),$$

where $H(\omega)$ is the frequency response function defined by

$$H(\omega) = \frac{1}{(\omega_0^2 - \omega^2)^2 + 2i\zeta\omega\omega_0}.$$

The response variances can be obtained by integrating the psd's over the entire frequency range. Thus,

$$\sigma_y^2 = \int_{-\infty}^{\infty} \Phi_y(\omega)\, d\omega,$$

$$\sigma_{\dot{y}}^2 = \int_{-\infty}^{\infty} \Phi_{\dot{y}}(\omega)\, d\omega, \qquad (17.81)$$

$$\sigma_{\ddot{y}}^2 = \int_{-\infty}^{\infty} \Phi_{\ddot{y}}(\omega)\, d\omega.$$

The integrations are performed numerically by using Simpson's rule. In so doing, the infinite limit is replaced by a finite limit equal to four times the natural frequency of the structure since the spectral density curves drop to almost zero for frequencies that slightly exceed the natural frequency. As the psd's of $\ddot{y}$, $\dot{y}$, and y are peaked near the natural frequency of the structure, most of the contributions to the integration intervals around the natural frequency are taken much finer.

The variation of mean wind velocity with height is given by the power law

$$V_Z/V_G = (Z/Z_G)^\alpha, \tag{17.82}$$

where V_Z is the mean wind velocity at altitude Z and V_G the gradient wind velocity at gradient height Z_G.

Objective Function

The objective function is the total volume of the structure and is given by

$$W(X) = \sum_{i=1}^{16} A_i L_i.$$

The notation used here is the same as that used for the earthquake design.

Frequency Constraint

The frequency constraint is also the same as that used for the earthquake design and takes the form

$$\omega^l \leqslant \omega_0(X).$$

Stress Constraint

The stresses in the members, due to the steady component of mean wind and the random dynamic component of wind, have to be considered separately. The force due to the former is

$$F_1 = \tfrac{1}{2}\rho A C_D V^2, \tag{17.83}$$

and the member stresses (also due to the steady component of wind) are

$$\{s_i\} = \frac{[B]\begin{Bmatrix} F_1 \\ 0 \\ \vdots \\ \vdots \\ 0 \end{Bmatrix}}{A_i} \qquad (i = 1, 2, \ldots, 16). \tag{17.84}$$

These stresses should be less than the yield stress σ_y in the problem formulation. The first part of the stress constraint thus becomes

$$\{s_i\} \leqslant \sigma_y \qquad (i = 1, 2, \ldots, 16). \tag{17.85}$$

The standard deviation of the acceleration (due to the gust component) is given by Eq. (17.81) and that of the force (also due to the gust component) by

$$\sigma_{F_1} = m\sigma_{\ddot{y}}, \tag{17.86}$$

Then the standard deviation of the stress in each member (due to the gust component) is

$$\{\sigma_{si}\} = \frac{[B]\begin{Bmatrix} \sigma_{F_1} \\ 0 \\ \vdots \\ \vdots \\ 0 \end{Bmatrix}}{A_i} \qquad (i = 1, 2, \ldots, 16). \tag{17.87}$$

The stresses caused by the gust, together with the mean stress in each member, should not exceed the yield stress. This condition is stated in probabilistic form because of the random nature of the stresses due to gust. The gust component is supposed to be a Gaussian process, and the first excursion failure probability of the random component of the stress in the i-th member exceeding $\sigma_y - s_i$, assuming independent crossings, is given by

$$q_i = 1 - \exp\left[-\frac{\omega T}{2\pi} \exp\{-(\sigma_y - |s_i|)^2/(2\sigma_{si}^2)\}\right]. \tag{17.88}$$

Here, the one-sided barrier-crossing probability used as the probability of the fluctuating stress crossing the other side of the barrier is almost always zero. Assuming full correlation between member stresses, we get the overall failure probability

$$q = \max_i \{q_i\}. \tag{17.89}$$

So the probability of failure constraint on the stress is given by

$$\max_i \left[1 - \exp\{-\frac{\omega T}{2\pi} \exp(-(\sigma_y - |s_i|)^2/(2\sigma_{si}^2))\}\right] \leqslant p_1. \tag{17.90}$$

Acceleration Constraint

The acceleration constraint is expressed as

$$\text{erfc}\left[a/(\sqrt{2}\sigma_{\ddot{y}})\right] \leqslant p_2 \tag{17.91}$$

which is in the same form as that for the earthquake design.

Optimization Problem Statement

The optimal design problem of the truss structure for an elevated water tank staging can be stated as: Find X to

$$\text{minimize } W(X) = \sum_{i=1}^{16} A_i L_i$$

subject to

$$g_1(X) = 1 - \omega^l/\omega_0(X) \geqslant 0,$$

$$g_{i+1}(X) = 1 - s_i/\sigma_y \geqslant 0 \quad (i = 1, 2, \ldots, 16),$$

$$g_{18}(X) = 1 - \max_{i=1}^{16} \left[1 - \exp\{-\frac{\omega T}{2\pi} \exp(-(\sigma_y - |s_i|)^2/(2\sigma_{si}^2))\}\right]/p_1 \geqslant 0,$$

$$g_{19}(X) = 1 - \text{erfc}\left[a/(\sqrt{2}\sigma_{\ddot{y}})\right]/p_2 \geqslant 0,$$

$$x_i \geqslant 0 \text{ (or } \epsilon_i) \quad (i = 1, 2, \ldots, 12),$$

where ϵ_i is the minimum allowable limit on the design variables.

The numerical computations for the optimization problem of the truss structure excited by wind loads is carried out for the following set of design data:

$$m = 5160 \text{ lb/in}^2, \quad H = 1080 \text{ in}, \quad L = 360 \text{ in}, \quad E = 30 \times 10^6 \text{ lb/in}^2,$$

$$\sigma_y = 30{,}000 \text{ lb/in}^2, \quad \omega^l = 5 \text{ rad/sec}, \quad p_1 = 10^{-5}, \quad p_2 = 10^{-4}, \quad a = 360 \text{ in/sec}^2,$$

$$T = 3600 \text{ sec}, \quad \alpha = 0.16, \quad \rho = 0.0024 \text{ slugs/ft}^3, \quad V_G = 140 \text{ ft/sec},$$

$$K = 0.005, \quad C_D = 2.0, \quad A = 400 \text{ ft}^2, \quad c/(m\omega_0) = 0.02.$$

The results are given in Table 17.5.

Table 17.5 Truss Structure (design for wind loads and earthquake acceleration)

	r	x_1	x_2	x_3	x_4	x_5	x_6	x_7	x_8	x_9	x_{10}	x_{11}	x_{12}	$W(X)$
Initial*	1.122	40	40	40	40	40	40	40	40	40	40	360	360	2,66,188
Optimal*	1.122×10^{-8}	14.8	39.6	15.8	40.1	20	15.5	1.47	9.85	14.3	0.4	389	372	1,05,983
Initial**	1.122	40	40	40	40	40	40	40	40	40	40	360	360	2,66,188
Optimal**	1.122×10^{-7}	12.2	43.7	15	36.6	18.9	15.7	4.6	9.3	13.9	2	380	385	1,07,243
Initial†	4.786	50	60	50	40	50	40	50	50	40	40	360	360	3,12,370
Optimal†	4.786×10^{-5}	10.5	48.3	14.6	21.6	19.4	17	20.2	9.7	13.8	8.5	371	374	1,21,400

* Design for wind loads without any constraint on member sizes
** Design for wind loads with constraints on minimum member sizes
† Earthquake design

We may draw the following conclusions from our study:

(i) From the results marked * in Table 17.5, it is observed that the areas of the horizontal members x_7 and x_{10} tend to vanish at the optimum point. In this event, the lower frequency and stress constraints are the active constraints. The results marked ** in Table 17.5 give a constraint on the minimum member sizes, namely, $x_i \geqslant 2$ for $i = 1, 2, \ldots, 10$. Here, the lower frequency and side constraints on the member area x_{10} are active constraints, whereas the stress constraint becomes inactive.

(ii) The optimal value of the objective function for wind loads, even for failure probability levels lower than those used for earthquake design, is smaller than the optimal values given in Table 17.4. This is due to the fact that the support structure is designed for the very severe earthquake El-Centro 1940. The optimal design for earthquakes that are approximately one-third the size of El-Centro 1940, with $G = 2$ (in/sec²)²/(rad/sec) and $p_1 = 10^{-4}$, is marked † in Table 17.5. The optimal design here is comparable to that for wind loads. If the design of the structure is to be so made as to safeguard the structure against both earthquakes and wind loads, the governing factor could be either the earthquake or the wind, depending on the relative seismicity and meteorological conditions of the location.

17.4.4 SHEET-STRINGER PANELS

Consider a sheet-stringer panel as shown in Fig. 17.7. If only a single panel between two stringers is considered instead of a multipanel system, then, according to Lin[18], the natural frequencies of such a system closely equal the limiting frequencies of the multipanel system. In each frequency band, the lowest frequency corresponds to the stringer-torsion mode and the highest frequency to the stringer-bending mode. Lin[19] considered the interaction of intermediate modes by means of a statistical argument and achieved close agreement between measured and computed stresses. Clarkson[20] has enumerated general principles for the design of fatigue-resistant structures. These principles specify conditions on the frequency which have to be satisfied by adjusting the panel parameters, such as stringer spacing and their torsional and bending stiffnesses. These conditions require (a) that the response levels be reduced by keeping these away from frequencies where the noise level is predominant and (b) a reasonably simplified analysis (based on the bounding frequencies) be made. In formulating our optimization problem for the sheet-stringer panels, the response analysis is made by taking into account the contribution from the two bounding frequencies in the first frequency band.

Considering the single panel between a pair of stringers and the two frames $x = 0$ and $x = a$, we get the governing differential equation for the plate motion as

$$D\left(\frac{\partial^4 w}{\partial x^4} + \frac{2\partial^4 w}{\partial^2 x\, \partial^2 y} + \frac{\partial^4 w}{\partial y^4}\right) + c\dot{w} + m_\mathrm{p}\ddot{w} = p(x, y, t), \tag{17.92}$$

where w is the transverse displacement of the plate, $D[= Et_\mathrm{s}^3/\{12(1 - v^2)\}]$ the bending rigidity of the plate, m_p the mass per unit area of the plate, c the damping coefficient per unit area of the plate, and $p(x, y, t)$ the random pressure field.

It is assumed that the sheet is simply-supported at the frames and the damping is viscous and small. On these assumptions, the normal mode approach can be used for analyzing the plate vibrations. The transverse deflection in terms of normal modes and generalized coordinates is expressed as

$$w(x, y, t) = \sum_{j=1}^{\infty} \sum_{k=1}^{\infty} \xi_j(x)\eta_k(y)q_{jk}(t). \tag{17.93}$$

Since the plate is simply-supported at the frames, the mode shapes along the X-direction may be assumed to be sinusoidal. Thus,

$$\xi_j(x) = \sin\frac{j\pi x}{a} \qquad (17.94)$$

In what follows, we assume that only the first mode in the X-direction makes a significant

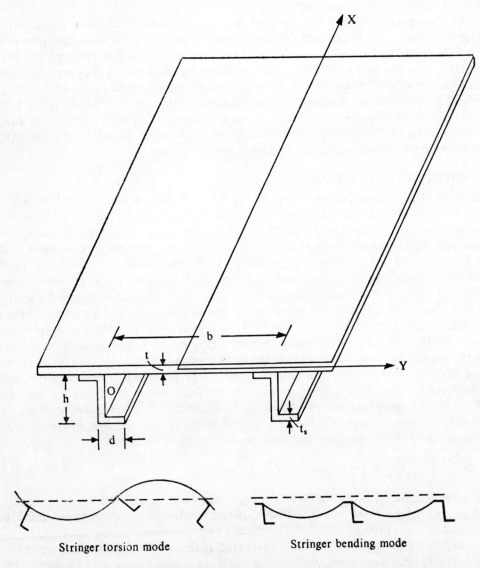

Stringer torsion mode   Stringer bending mode

Fig. 17.7   Sheet-Stringer Combination and Normal Modes Corresponding to Bounding Frequencies.

contribution to the response. This is reasonable since the jet noise is supposed to be fully correlated along the X-direction. Hence, the joint acceptance vanishes for the even-numbered modes in the X-direction and progressively decreases in magnitude for the odd-numbered modes (see Bozich[21]).

The governing differential equation in generalized coordinates can be expressed as

$$M_j\ddot{q}_j(t) + C_j\dot{q}_j(t) + K_jq_j = p_j(t), \tag{17.95}$$

where

$$w(x, y, t) = \sum_{j=1}^{\infty} f_j(x, y)q_j(t), \qquad f_j(x, y) = \sin\frac{\pi x}{a}\eta_j(y),$$

and M_j, K_j, C_j, and p_j are, respectively, the generalized mass, stiffness, damping, and force.

Following Lin[18], we get both the frequency equation and the mode shape of the plate for the stringer-torsion mode and for the stringer-bending mode as

$$\cosh\frac{K_1b}{2}(K_2\sin\frac{K_2b}{2} + 2DK_2^2\cos\frac{K_2b}{2}) + \cos\frac{K_2b}{2}(\beta K_1\sinh\frac{K_1b}{2} + 2DK_1^2\cosh\frac{K_1b}{2}) = 0, \tag{17.96}$$

$$\eta_t(y) = \cosh K_1y - \alpha\cosh K_2y, \tag{17.97}$$

$$K_1'\sinh\frac{K_1'b}{2}(\gamma\cos\frac{K_2'b}{2} - 2DK_2'^3\sinh\frac{K_2'b}{2})$$
$$+ K_2'\sinh\frac{K_2'b}{2}(\gamma\cosh\frac{K_1'b}{2} - 2DK_1'^3\sinh\frac{K_1'b}{2}) = 0, \tag{17.98}$$

$$\eta_b(y) = \cosh K_1'y + \alpha'\cos K_2'y. \tag{17.99}$$

Jet Noise Pressure Field

The cross psd of the jet noise pressure field is assumed to be of the form

$$\Phi_{pp}(x_1, y_1; x_2, y_2; \omega) = \Phi_p(\omega)e(x_1, y_1; x_2, y_2; \omega), \tag{17.100}$$

where $\Phi_p(\omega)$ is the psd of the normalizing homogeneous pressure field and $e(x_1, y_1; x_2, y_2; \omega)$ is in the form of a correlation coefficient. The jet noise is assumed to be a weakly stationary Gaussian random process. It is supposed to be fully correlated in the X-direction and homogeneous in the Y-direction. The psd of the jet noise is taken to be

$$\Phi_{pp}(x_1, y_1; x_2, y_2; \omega) = \Phi_p(\omega)\cos[\frac{\omega\cos\theta}{v}(y_1 - y_2)], \tag{17.101}$$

where ω is the jet noise frequency, v the speed of sound in air, and θ the angle of the predominant direction of the jet noise with the Y-axis. The form of this function is the same as that used by Wallace[22].

It can be shown that the psd of the transverse displacement is

$$\Phi_{ww}(x_1, y_1; x_2, y_2; \omega) = \sum_{j=1}^{\infty}\sum_{k=1}^{\infty} f_j\frac{f_j(x_1, y_1)f_k(x_2, y_2)}{Z_j(\omega)Z_k^*(\omega)}I_{jk}(\omega), \tag{17.102}$$

where $Z_j(\omega)$ $[=1/(-M_j\omega^2 + iC_j\omega + K_j)]$ is the complex impedance function in the j-th mode, * denotes the complex conjugate, and

$$I_{jk}(\omega) = \Phi_p(\omega)\int_A\int_A f_j(x_1, y_1)f_k(x_2, y_2)e(x_1, y_1; x_2, y_2; \omega)\,dA_1\,dA_2. \tag{17.103}$$

The psd's of the stresses at different locations can be obtained as

$$\Phi_{sxsx}(x_1, y_1; x_2, y_2; \omega) = \sum_{j=1}^{\infty} \sum_{k=1}^{\infty} \frac{g_{jx}(x_1, y_1)g_{kx}(x_2, y_2)}{Z_j(\omega)Z_k^*(\omega)} I_{jk}(\omega), \tag{17.104}$$

$$\Phi_{sxsy}(x_1, y_1; x_2, y_2; \omega) = \sum_{j=1}^{\infty} \sum_{k=1}^{\infty} \frac{g_{jy}(x_1, y_1)g_{ky}(x_2, y_2)}{Z_j(\omega)Z_k^*(\omega)} J_{jk}(\omega), \tag{17.105}$$

$$\Phi_{sxsy}(x_1, y_1; x_2, y_2; \omega) = \sum_{j=1}^{\infty} \sum_{k=1}^{\infty} \frac{g_{jx}(x_1, y_1)g_{ky}(x_2, y_2)}{Z_j(\omega)Z_k^*(\omega)} I_{jk}(\omega), \tag{17.106}$$

where

$$g_{jx} = -\frac{Et_s}{1 - v^2}\left(\frac{\partial^2 f_j}{\partial x^2} + v\frac{\partial^2 f_j}{\partial y^2}\right),$$

$$g_{jy} = -\frac{Et_s}{1 - v^2}\left(\frac{\partial^2 f_j}{\partial y^2} + v\frac{\partial^2 f_j}{\partial x^2}\right).$$

From the psd's, the variance and the correlation coefficients between stresses at different locations can be obtained by integration as

$$\sigma_{sx}^2(x, y) = \int_{-\infty}^{\infty} \Phi_{sxsx}(x, y; x, y; \omega)\,d\omega, \tag{17.107}$$

$$\sigma_{sy}^2(x, y) = \int_{-\infty}^{\infty} \Phi_{sysy}(x, y; x, y; \omega)\,d\omega, \tag{17.108}$$

$$\rho_{sxsx}(x_1, y_1; x_2, y_2) = \frac{\int_{-\infty}^{\infty} \Phi_{sxsx}(x_1, y_1; x_2, y_2; \omega)\,d\omega}{\sigma_{sx}(x_1, y_1)\sigma_{sx}(x_2, y_2)}, \tag{17.109}$$

$$\rho_{sysy}(x_1, y_1; x_2, y_2) = \frac{\int_{-\infty}^{\infty} \Phi_{sysy}(x_1, y_1; x_2, y_2; \omega)\,d\omega}{\sigma_{sy}(x_1, y_1)\sigma_{sy}(x_2, y_2)}, \tag{17.110}$$

$$\rho_{sxsy}(x_1, y_1; x_2, y_2) = \frac{\int_{-\infty}^{\infty} \Phi_{sxsy}(x_1, y_1; x_2, y_2; \omega)\,d\omega}{\sigma_{sx}(x_1, y_1)\sigma_{sy}(x_2, y_2)}. \tag{17.111}$$

MINIMUM WEIGHT DESIGN OF SHEET-STRINGER PANELS—PROBLEM FORMULATION

Objective Function

While designing sheet-stringer panels, the distance a between the frames is specified and the density of the material is constant. The objective function is therefore taken to be the area of cross-section of the sheet-stringer panel. The design variables in the problem are the stringer dimensions—spacing b, web height h, flange width d, thickness t_s—and the thickness of the plate t. Thus, each possible design is fully defined by the components of the vector

$$X^T = \{x_1, x_2, x_3, x_4, x_5\}^T = \{b, h, d, t, t_s\}^T, \tag{17.112}$$

and hence the objective function is

$$W(X) = bt + t_s(h + 2d). \tag{17.113}$$

Frequency Constraints

In the stringer-torsion mode, the adjacent panels vibrate out of phase, whereas in the stringer-bending mode they vibrate in phase. In the frequency range of 400–1000 Hz, the jet noise pressure correlation is high and positive up to about 2 ft. If the stringer pitch is approximately 6 in, at least four panels have pressures in phase. Thus, the stringer-torsion mode does not get appreciably excited by the jet noise as compared with the stringer-bending mode. To reduce the response, the pressure spectrum should be so used that the frequency of the sheet corresponding to the stringer-bending mode is well above the frequency of the maximum sound energy. Such a condition can be satisfied if the lower bending frequency itself is above the frequency of the maximum sound energy. This requirement takes the form of the frequency constraint

$$\omega_t(X) \geqslant \omega^l, \tag{17.114}$$

where $\omega_t(X)$ is the frequency of the plate vibrations corresponding to the stringer-torsion mode. To ensure that the stringer-torsion frequency does not in any way exceed the stringer-bending frequency, the second frequency constraint is expressed as

$$\omega_b(X) \geqslant \omega_t(X), \tag{17.115}$$

where $\omega_b(X)$ is the frequency of the plate vibrations corresponding to the stringer-bending mode.

Fatigue Damage Constraint

The critical stresses (see Clarkson[23]) are $s_x(a/2, b/2)$, $s_y(a/2, b/2)$, and $s_y(a/2, 0)$. These stresses are Gaussian since the excitation is Gaussian. The expected rate of fatigue damage is calculated on the basis of these stresses by using the Palmgren-Miner cumulative damage rule. For the crossing rate of the stress cycles, an average frequency is used to arrive at the fatigue damage constraint which is expressed as

$$\frac{1}{2\pi\beta}\left(\frac{\omega_t + \omega_b}{2}\right)[\sqrt{2}\sigma_{sx}(a/2, b/2)]^\alpha \Gamma\left(1 + \frac{\alpha}{2}\right) \leqslant D_1, \tag{17.116}$$

$$\frac{1}{2\pi\beta}\left(\frac{\omega_t + \omega_b}{2}\right)[\sqrt{2}\sigma_{sy}(a/2, b/2)]^\alpha \Gamma\left(1 + \frac{\alpha}{2}\right) \leqslant D_1, \tag{17.117}$$

$$\frac{1}{2\pi\beta}\left(\frac{\omega_t + \omega_b}{2}\right)[\sqrt{2}\sigma_{sy}(a/2, 0)]^\alpha \Gamma\left(1 + \frac{\alpha}{2}\right) \leqslant D_1, \tag{17.118}$$

where α, β are the constants in the fatigue law given by $NS^\alpha = \beta$, D_1 is the specified damage level, and $\Gamma(\ldots)$ the gamma function.

Stress Constraint

When the critical stresses exceed the yield stress, failure occurs. The probability of failure based on the stress response can be expressed as either individual stresses at the critical locations or an overall probability of failure. The stress constraint can be defined in one of the following ways:

(i) Failure based on individual stresses given by

$$\text{erfc}\left[\frac{\sigma_y}{\sqrt{2}\sigma_{sx}(a/2, b/2)}\right] \leqslant p_1, \tag{17.119}$$

$$\text{erfc}\left[\frac{\sigma_y}{\sqrt{2}\sigma_{sy}(a/2,\ b/2)}\right] \leqslant p_1, \tag{17.120}$$

$$\text{erfc}\left[\frac{\sigma_y}{\sqrt{2}\sigma_{sy}(a/2,\ 0)}\right] \leqslant p_1. \tag{17.121}$$

(ii) Failure based on the upper bound given by the inequality (see Shinozuka[16])

$$\text{erfc}\left[\frac{\sigma_y}{\sqrt{2}\sigma_{sx}(a/2,\ b/2)}\right] + \text{erfc}\left[\frac{\sigma_y}{\sqrt{2}\sigma_{sy}(a/2,\ b/2)}\right] + \text{erfc}\left[\frac{\sigma_y}{\sqrt{2}\sigma_{sy}(a/2,\ 0)}\right] \leqslant p_2. \tag{17.122}$$

(iii) Failure based on inequality (17.8) given by

$$\text{erfc}\left[\frac{\sigma_y}{\sqrt{2}\sigma_{sy}(a/2,\ 0)}\right] + C_1 \text{erfc}\left[\frac{\sigma_y}{\sqrt{2}\sigma_{sy}(a/2,\ b/2)}\right] + C_2 \text{erfc}\left[\frac{\sigma_y}{\sqrt{2}\sigma_{sx}(a/2,\ b/2)}\right] \leqslant p_2, \tag{17.123}$$

where

$$C_1 = 1 - \frac{\text{erfc}(a_1)}{\text{erfc}\left[\dfrac{\sigma_y}{\sqrt{2}\sigma_{sy}(a/2,\ b/2)}\right]},$$

$$a_1 = \max\left[\frac{\sigma_y}{\sqrt{2}\sigma_{sy}(a/2,\ b/2)},\ \frac{\sigma_y\sigma_{sy}(a/2,\ b/2)}{\sqrt{2}\sigma_{sy}^2(a/2,\ 0)|\rho_{sysy}(a/2,\ b/2;\ a/2,\ 0)|}\right],$$

$$C_2 = \min\left\{1 - \frac{\text{erfc}(a_2)}{\text{erfc}\left[\dfrac{\sigma_y}{\sqrt{2}\sigma_{sx}(a/2,\ b/2)}\right]},\ 1 - \frac{\text{erfc}(a_3)}{\text{erfc}\left[\dfrac{\sigma_y}{\sqrt{2}\sigma_{sx}(a/2,\ b/2)}\right]}\right\},$$

$$a_2 = \max\left[\frac{\sigma_y}{\sqrt{2}\sigma_{sx}(a/2,\ b/2)},\ \frac{\sigma_y\sigma_{sx}(a/2,\ b/2)}{\sqrt{2}\sigma_{sy}^2(a_2,\ 0)|\rho_{sxsy}(a/2,\ b/2;\ a/2,\ 0)|}\right],$$

$$a_3 = \max\left[\frac{\sigma_y}{\sqrt{2}\sigma_{sx}(a/2,\ b/2)},\ \frac{\sigma_y\sigma_{sx}(a/2,\ b/2)}{\sqrt{2}\sigma_{sy}^2(a/2,\ b/2)|\rho_{sxsy}(a/2,\ b/2;\ a/2,\ b/2)|}\right].$$

In these constraints, erfc (...) denotes the complementary error function and σ_y the yield stress which is assumed to be deterministic.

Side Constraints

We have assumed that the mode shape in the X-direction is sinusoidal. This assumption is valid only if the frame spacing a is greater by one-and-a-half times the stringer spacing b (see Vanmarcke[12]), which is a constraint expressed as

$$a \geqslant 1.5b. \tag{17.124}$$

The constraints on the thickness of sheet and stringer material are in the usual range of thickness of aluminium sheet, and are given by

$$0.02 \leqslant t \leqslant 0.2, \tag{17.125}$$

$$0.02 \leqslant t_s \leqslant 0.2. \tag{17.126}$$

When formulating the problem, we should ensure that there is no tendency for the stringer spacing to reduce indefinitely during optimization. This constraint is expressed as

$$b \geqslant 2h. \tag{17.127}$$

The side constraints on the minimum dimensions of the web height and the flange width of the stringer are specified as

$$h \geqslant 0.5, \tag{17.128}$$

$$d \geqslant 0.5. \tag{17.129}$$

It may be noted that the non-negativity restrictions of the design variables are included in the side constraints, and are defined by

$$x_i \geqslant 0 \quad (i = 1, 2, \ldots, 5). \tag{17.130}$$

Optimization Problem Statement

The minimum weight design problem for the sheet-stringer panels subjected to jet noise can be stated as: Find X^* to

$$\text{minimize } W(X) = bt + t_s(h + 2d) \tag{17.131}$$

subject to

$$g_1(X) = 1 - \omega^l/\omega_t(X) \geqslant 0,$$

$$g_2(X) = 1 - \omega_t(X)/\omega_b(X) \geqslant 0,$$

$$g_3(X) = 1 - \frac{1}{2\pi\beta D_1}(\frac{\omega_t + \omega_b}{2})[\sqrt{2}\sigma_{sx}(a/2, b/2)]^\alpha \Gamma(1 + \frac{\alpha}{2}) \geqslant 0,$$

$$g_4(X) = 1 - \frac{1}{2\pi\beta D_1}(\frac{\omega_t + \omega_b}{2})[\sqrt{2}\sigma_{sy}(a/2, b/2)]^\alpha \Gamma(1 + \frac{\alpha}{2}) \geqslant 0,$$

$$g_5(X) = 1 - \frac{1}{2\pi\beta D_1}(\frac{\omega_t + \omega_b}{2})[\sqrt{2}\sigma_{sy}(a/2, 0)]^\alpha \Gamma(1 + \frac{\alpha}{2}) \geqslant 0,$$

$$g_6(X) = 1 - \frac{\text{erfc}\,[\dfrac{\sigma_y}{\sqrt{2}\sigma_{sx}(a/2,\ b/2)}]}{p_1} \geqslant 0,$$

$$g_7(X) = 1 - \frac{\text{erfc}\,[\dfrac{\sigma_y}{\sqrt{2}\sigma_{sy}(a/2,\ b/2)}]}{p_1} \geqslant 0, \tag{17.132a}$$

$$g_8(X) = 1 - \frac{\text{erfc}\,[\dfrac{\sigma_y}{\sqrt{2}\sigma_{sy}(a/2,\ 0)}]}{p_1} \geqslant 0,$$

$$g_9(X) = 1 - \frac{1.5b}{a} \geqslant 0,$$

$$g_{10}(X) = 1 - \frac{t}{0.2} \geqslant 0,$$

$$g_{11}(X) = 1 - \frac{t_s}{0.2} \geqslant 0,$$

$$g_{12}(X) = 1 - \frac{0.02}{t} \geqslant 0,$$

$$g_{13}(X) = 1 - \frac{0.02}{t_s} \geqslant 0,$$

$$g_{14}(X) = 1 - \frac{2h}{b} \geqslant 0,$$

$$g_{15}(X) = 1 - \frac{0.5}{h} \geqslant 0,$$

$$g_{16}(X) = 1 - \frac{0.5}{d} \geqslant 0,$$

$$x_i \geqslant 0 \qquad (i = 1, 2, \ldots, 5).$$

(17.132b)

The foregoing formulation can be modified if the stress constraint is expressed as (17.122) or (17.123). In that event, the constraints g_6, g_7, g_8 are replaced by a single constraint of the form

$$g_6(X) = 1 - \frac{\text{erfc}\,[\frac{\sigma_y}{\sqrt{2}\sigma_{sx}(a/2,b/2)}] + \text{erfc}\,[\frac{\sigma_y}{\sqrt{2}\sigma_{sy}(a/2,b/2)}] + \text{erfc}\,[\frac{\sigma_y}{\sqrt{2}\sigma_{sy}(a/2,0)}]}{p_2} \geqslant 0$$

(17.133)

or

$$g_6(X) = 1 - \frac{\text{erfc}\,[\frac{\sigma_y}{\sqrt{2}\sigma_{sy}(a/2,0)}] + C_1\,\text{erfc}\,[\frac{\sigma_y}{\sqrt{2}\sigma_{sy}(a/2,b/2)}] + C_2\,\text{erfc}\,[\frac{\sigma_y}{\sqrt{2}\sigma_{sx}(a/2,b/2)}]}{p_2} \geqslant 0$$

(17.134)

and the constraints g_9–g_{16} are numbered consecutively from g_7–g_{14}.

The numerical solution for the optimization problem is obtained for two sets of design data which differ in the levels of the white noise psd of the jet noise pressures $\Phi_p(\omega)$ and the limiting value specified on the frequency corresponding to the stringer-torsion mode. These sets of data are:

$$\Phi_p(\omega) = 0.41 \times 10^{-4}\,(\text{lb/in}^2)^2/(\text{rad/sec}), \qquad \omega^l = 1884\,\text{rad/sec};$$

(17.135)

$$\Phi_p(\omega) = 0.7 \times 10^{-5}\,(\text{lb/in}^2)^2/(\text{rad/sec}), \qquad \omega^l = 942\,\text{rad/sec}.$$

(17.136)

The remaining design specifications, which are the same in both sets (17.135) and (17.136), are:

$$a = 20\,\text{in}, \qquad \zeta = 0.02, \qquad \sigma_y = 20{,}000\,\text{lb/in}^2, \qquad \alpha = 6, \qquad \beta = 6.4 \times 10^{31},$$
$$v = 1100\,\text{ft/sec}, \qquad \theta = \pi/4\,\text{rad}, \qquad \rho = 0.1\,\text{lb/in}^3.$$

(17.137)

Tables 17.6 and 17.7 provide the optimum design for a given failure probability and damage level [as expressed in inequalities (17.119)–(17.123)]. Some of these results are plotted in Fig. 17.8.

The minimum weight design problem for sheet-stringer panels subjected to jet noise excitation, which we have formulated within the reliability framework, was reduced to an NLP problem and solved on a digital computer by employing the unconstrained minimization technique. We may draw the following conclusions from our study.

(i) The optimum design is sensitive to the constraints on a specified probability of failure and fatigue damage level. A comparison of the data marked ** and *** in Table 17.6 demonstrates a 14.2 per cent increase in weight for a decrease in probability of failure from 10^{-4} to 10^{-5}; the data marked * shows that the stress constraints are not active. A comparison of the results marked *** and † in Table 17.6 shows a 17 per cent increase in weight for a change in damage level from 5×10^{-6} to 1×10^{-6}.

(ii) The constraints on stresses were imposed in three different ways [see inequalities (17.119)–(17.123)]. The results indicate that the optimum design is insensitive to the manner in which the constraints are applied. This is particularly true when the stress constraint is inactive at the optimum. In other cases, the optimum design is sensitive because the stress at one location is considerably more than the stresses at other locations.

Table 17.6 Optimum Design for Sheet-Stringer Panel

Given

$$\Phi_p(\omega) = 0.41 \times 10^{-4} \ (\text{lb/in}^2)^2/(\text{rad/sec}), \qquad \omega^l = 1884 \ \text{rad/sec}.$$

	r	x_1	x_2	x_3	x_4	x_5	$W(X)$	$W(X) + r\Sigma \dfrac{1}{g_j(X)}$	$\dfrac{W(X)}{b}$
Initial*	1.384×10^{-2}	5	2.4	0.6	0.12	0.08	0.888	2.367	0.176
Optimal*	1.384×10^{-5}	3.621	1.801	0.516	0.061	0.0597	0.390	0.395	0.108
Initial**	1.384×10^{-2}	5	2.4	0.6	0.12	0.08	0.888	2.367	0.176
Optimal**	1.384×10^{-6}	3.536	1.765	0.516	0.0604	0.064	0.393	0.394	0.111
Initial***	1.384×10^{-2}	5	2.4	0.6	0.12	0.08	0.888	2.367	0.176
Optimal***	1.384×10^{-8}	3.706	1.853	0.523	0.0676	0.069	0.451	0.451	0.121
Initial†	1.384×10^{-2}	5	2.4	0.6	0.12	0.08	0.888	2.367	0.176
Optimal†	1.384×10^{-5}	4.051	2.018	0.522	0.0779	0.0694	0.528	0.551	0.130
Initial‡	1.51×10^{-2}	5	2.4	0.6	0.12	0.08	0.888	2.367	0.176
Optimal‡	1.51×10^{-6}	3.546	1.770	0.518	0.0606	0.0634	0.393	0.395	0.110

*$p_2 = 10^{-3}$, $D_1 = 5 \times 10^{-6}$, active constraints g_5, g_{12} (stringer spacing)
**$p_2 = 10^{-4}$, $D_1 = 5 \times 10^{-6}$, active constraints g_5, g_6, g_{12} (stringer spacing)
***$p_2 = 10^{-5}$, $D_1 = 5 \times 10^{-6}$, active constraints g_6, g_{12} (stringer spacing)
†$p_2 = 10^{-5}$, $D_1 = 10^{-6}$, active constraints g_5, g_{12} (stringer spacing)
‡$p_2 = 10^{-4}$, $D_1 = 5 \times 10^{-6}$, active constraints g_8, g_{14} (stringer spacing)

Table 17.7 Optimum Design for Sheet-Stringer Panel

Given

$$\Phi_p(\omega) = 0.7 \times 10^{-5} \ (\text{lb/in}^2)^2/(\text{rad/sec}), \qquad \omega^l = 942 \ \text{rad/sec}.$$

	r	x_1	x_2	x_3	x_4	x_5	$W(X)$	$W(X) + r\Sigma \dfrac{1}{g_j(X)}$	$\dfrac{W(X)}{b}$
Initial*	1.703×10^{-2}	8	2	1	0.08	0.08	0.96	1.923	0.12
Optimal*	1.703×10^{-7}	2.897	1.447	0.501	0.035	0.0361	0.190	0.191	0.0655
Initial**	1.649×10^{-2}	8	2	1	0.08	0.08	0.96	1.923	0.12
Optimal**	1.649×10^{-7}	3.044	1.521	0.501	0.0361	0.0321	0.091	0.191	0.0655

*$p_1 = 10^{-3}$, $D_1 = 10^{-6}$, active constraints g_5, g_{14}, g_{16}
**$p_2 = 10^{-3}$, $D_1 = 10^{-6}$, active constraints g_5, g_{12}, g_{14} (stringer spacing)

(iii) The side constraint on the stringer spacing is active in all cases. This is expected for a single panel because the reduction in stringer spacing shortens the area of the sheet and the size of the stringers. For a structural system, such as a fuselage, the minimum weight of a panel does not always imply minimum weight of the total system. It is therefore appropriate to consider the area of a single panel per unit length of the stringer spacing as the objective function. That is,

$$W(X) = t + t_s(h + 2d)/d. \tag{17.138}$$

The results for the objective function given by Eq. (17.138) and the design data given by

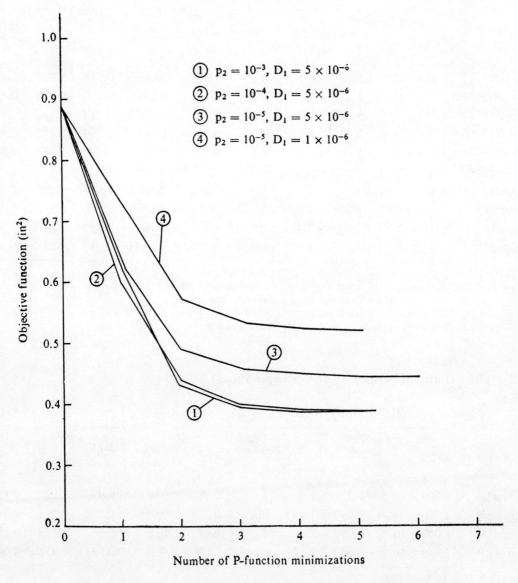

Fig. 17.8 Sequence of Unconstrained Optima (Sheet-stringer panel).

(17.136) and (17.137) are listed in Table 17.8. A comparison of the results in this table with those in Table 17.7 shows that the formulation based on (17.138) leads to a 15.4 per cent reduction in weight.

Table 17.8   Optimum Design for Sheet-Stringer Panel

Given

$$\Phi_p(\omega) = 0.7 \times 10^{-5} \ (\text{lb/in}^2)^2/(\text{rad/sec}), \qquad \omega^l = 942 \ \text{rad/sec}.$$

	r	x_1	x_2	x_3	x_4	x_5	$W(X)$	$W(X) + r\Sigma\dfrac{1}{g_j(X)}$
Initial*	2.061×10^{-3}	8	2	1	0.08	0.08	0.12	0.24
Optimal*	2.061×10^{-9}	5.954	1.702	1.06	0.0403	0.023	0.05505	0.05506
Initial**	2.044×10^{-3}	8	2	1	0.08	0.08	0.12	0.2403
Optimal**	2.044×10^{-9}	6.70	1.822	1.276	0.042	0.02	0.05518	0.05518

* $p_2 = 10^{-3}$, $D_1 = 10^{-6}$, active constraint g_5
** $p_2 = 10^{-3}$, $D_1 = 10^{-6}$, active constraints g_5, g_{13}

REFERENCES

1   Ang, A. H. S., Structural safety: a literature review, *Proc. ASCE, J. Struct. Div.*, **98**, ST4, 845–54, 1972.

2   Shinozuka, M., Safety against dynamic forces, *Proc. ASCE, J. Struct. Div.*, **100**, ST9, 1821–26, 1974.

3   Yang, J. N. and Heer, E., Reliability of randomly excited structures, *AIAA J.*, **9**, 7, 1262–68, 1971.

4   Yang, J. N. and Shinozuka, M., On the first excursion probability in stationary narrow band random vibration, *J. Appl. Mech.*, **38**, 4, 1017–22, 1971.

5   Racicot, R. L., Random vibration analysis: application to wind loaded structures, Ph.D. Thesis, Department of Civil Engineering, Case Western Reserve University, Ohio, 1969.

6   Nigam, N. C., Structural optimization in random vibration environment, *AIAA J.*, **10**, 4, 551–53, 1972.

7   Narayanan, S., Structural optimization in random vibration environment, Ph.D. Thesis, Department of Aeronautical Engineering, IIT, Kanpur, 1975.

8   Fiacco, A. V. and McCormick, G. P., The sequential unconstrained minimization technique for nonlinear programming: a primal dual method, *Management Sc.*, **10**, 2, 360–66, 1964.

9   Fletcher, R. and Powell, M. J. D., A rapidly convergent descent method for minimization, *Computer J.*, **6**, 2, 163–68, 1963.

10   Vanmarcke, E. H., Matrix formulation of reliability analysis and reliability based design, National Symp. on Computer Structural Analysis and Design, George Washington University, 1972.

11 Miner, M. A., Cumulative damage in fatigue, *J. Appl. Mech.*, **12**, 3, 159–64, 1945.

12 Vanmarcke, E. H., First passage and other failure criteria in narrow band random vibration: a discrete state approach, Ph.D. Thesis, Department of Civil Engineering, M. I. T., Boston, 1969.

13 Pestel, E. and Leckie, F., Matrix Methods in Elastomechanics, McGraw-Hill, New York, 1963.

14 Housner, G. W., The dynamic behaviour of water tanks, *Bul. Seis. Soc. Am.*, **53**, 2, 381–87, 1963.

15 Housner, G. W. and Jennings, P. C., Generation of artificial earthquakes, *Proc. ASCE, J. Engg. Mech. Div.*, **90**, EM1, 113–50, 1964.

16 Shinozuka, M., Probability of structural failure under random loading, *Proc. ASCE, J. Engg. Mech. Div.*, **90**, EM5, 147–70, 1964.

17 Davenport, A. G., The application of statistical concepts to wind loading of structures, *Proc. Inst. of Civil Engrs.*, **19**, 8, 449–73, 1961.

18 Lin, Y. K., Free vibration of continuous skin-stringer panels, *J. Appl. Mech.*, **27**, 4, 669–76, 1960.

19 Lin, Y. K., Stresses in continuous skin-stiffener panels under random loading, *J. Aero. Space Sc.*, **29**, 1, 67–75, 1962.

20 Clarkson, B. L., Design of fatigue resistant structures. *In* Richard, E. J. and Mead, D. J. (eds.) Noise and Acoustic Fatigue in Aeronautics, Wiley, New York, 1968.

21 Bozich, O. J., Spatial correlation in acoustic structural coupling, *J. Acous. Soc. Am.*, **36**, 1, 52–58, 1964.

22 Wallace, C. E., Response and fatigue life of acoustically excited sandwich panels. *In* Trapp, W. J. and Forney, D. M. (eds.) Acoustic Fatigue in Aerospace Structures, Syracuse University Press, Syracuse, 1965.

23 Clarkson, B. L., Stresses in skin panels subjected to random acoustic loading, *Aeronautical J. Roy. Aero. Soc.*, **72**, 2, 1000–1010, 1968.

Index